William Stark

The Works of the Late William Stark, M.D.

Consisting of clinical and anatomical observations, with experiments, dietetical and statical, revised and published from his original mss. by James Carmichael Smyth

William Stark

The Works of the Late William Stark, M.D.
Consisting of clinical and anatomical observations, with experiments, dietetical and statical, revised and published from his original mss. by James Carmichael Smyth

ISBN/EAN: 9783337042554

Printed in Europe, USA, Canada, Australia, Japan

Cover: Foto ©berggeist007 / pixelio.de

THE

WORKS

OF THE LATE

WILLIAM STARK, M.D.

CONSISTING OF

CLINICAL AND ANATOMICAL OBSERVATIONS,

WITH

EXPERIMENTS,

DIETETICAL AND STATICAL,

REVISED AND PUBLISHED FROM HIS ORIGINAL MSS.

BY

JAMES CARMICHAEL SMYTH, M.D. F.R.S.
PHYSICIAN EXTRAORDINARY TO HIS MAJESTY.

LONDON:

PRINTED FOR J. JOHNSON, No. 72, ST. PAUL'S CHURCH-YARD.

M.DCC.LXXXVIII.

TO

The Hon. THOMAS FITZMAURICE.

SIR,

AS mankind are commonly desirous of knowing the persons to whom they are indebted, it is proper for me to inform them, that if any useful knowledge is contained in the following pages, they owe it chiefly to you. You distinguished, at an early period of life, the talents and abilities of the Author; you honoured him with your friendship, encouraged him by your protection, and your regard for his memory has preserved his works from oblivion, which, otherwise, would have perished with himself. You at first committed them to the care of a Gentleman, very capable to have done them justice, had not bad health and a variety of avocations, prevented him. For myself, I can only say, that I have executed, with all the zeal and ardour of friendship, a task which I formerly undertook at the request of the Author, and to the performance of which I felt myself urged by those sentiments which are so happily expressed in the energetic and manly language of the firſt of hiſtorians. " Non hoc præcipuum " amicorum munus eſt, proſequi defunctum ignavo queſtu; ſed quæ " voluerit meminiſſe; quæ mandaverit exſequi*."

Permit me, Sir, to acknowledge likewiſe, the pleaſure which this occaſion affords me, of publickly declaring the great regard and eſteem, with which I have the honour to be,

S I R,

Your moſt obedient,

And moſt humble ſervant,

JAMES CARMICHAEL SMYTH.

* Tacit. Annal. lib. ii. cap. lxxi.

THE

PREFACE BY THE EDITOR.

ALTHOUGH the importance and ſcarcity of original Obſervations in Phyſic are well known, yet I am ready to confeſs, that neither the merits or originality of the preſent work, nor even my knowledge of the accuracy and candour of the Author, would have been ſufficient inducements with me to become the Editor, had I not felt a ſtrong deſire to comply with the requiſition of a friend, now no more, and a wiſh to preſerve to his memory, the fame he ſo juſtly deſerves, and which he ſo dearly earned. It is evident that I was influenced by no pecuniary motive, as any emolument ariſing from the ſale of the book is deſtined to his family; and I could expect but little reputation from publiſhing the works of another, compared with the time and trouble which I muſt neceſſarily devote to it.

As Editor, my chief object has been perſpicuity; and to effect this, I have taken conſiderable liberties both with the arrangement and language of the Author; adhering, however, with the moſt ſcrupulous exactneſs, to facts, and, wherever I could, retaining his own language, which, though ſometimes obſcure, is commonly expreſſive and manly. I am extremely ſenſible that this mixture of of compoſition muſt affect the unity and ſmoothneſs of the ſtile, but, in works of ſcience, it is clearneſs and preciſion, more than elegance, that is wanted.

The different parts into which the Clinical and Anatomical Obſervations are diſtributed, though equally correct, are by no means equally complete, or equally uſeful. It was, at firſt, my intention, to have endeavoured to ſupply what appeared to me defective, and to have ſubjoined a comment to each part, in the manner I formerly did to the one publiſhed in the Medical Communications, but I ſoon found that I had neither time or leiſure, at preſent, for ſuch an undertaking. I ſhall therefore only obſerve in general, that from the Clinical and Anatomical Obſervations of our Author, the young may receive much uſeful information, and that even the more experienced may find ſomething to learn.

His Experiments on Diet are the firſt, and will probably long remain the only Experiments of the kind. It will poſſibly be objected to them, that they are not ſufficiently numerous or varied to admit of any concluſions, but I would adviſe thoſe who may ſtart ſuch objections, to reflect, that all inductions from experience, are, at beſt, only greater or leſs degrees of probability, and that if one Experiment did not afford ſome probability, twenty Experiments could not eſtabliſh any. But Dr. Stark's Experiments will be looked upon by all true lovers of ſcience, in a much higher point of view. They will be conſidered as the corner-ſtone of a great building, to be finiſhed at ſome after-period of time, when men ſhall be found of equal fortitude, perſeverance

ſeverance and ſelf-denial with our Author, actuated by a ſimilar zeal for the promoting of uſeful knowledge.

Having ſaid thus much of the Work, it may be expected that I ſhould ſay ſomething of the Author himſelf. This I do with a ſingular pleaſure, as it is tracing back in my remembrance, the image of a friend I eſteemed and valued, and to whom I am certain it would have afforded a ſingular ſatisfaction, had he known that I would have been the publiſher of his works, and the hiſtorian of his life.

Dr. Stark's father, as I have been told, was a native of Ireland, his mother of Scotland, he himſelf was born at Birmingham. This might be conſidered as a whimſical anecdote only, and ſcarcely deſerving notice, did we not every day ſee the characters and conduct of men influenced by ſuch trifling and accidental circumſtances, and therefore it may not be unreaſonable to ſuppoſe, that even this had ſome effect in expanding the natural liberality of his mind, and freeing it from all thoſe little local and confined prejudices, which too often diſgrace characters otherwiſe reſpectable. But, wherever his cradle was rocked, he was educated at Glaſgow, and there, under Drs. Adam Smith, Black, Reid, &c. he firſt learnt the rudiments of philoſophy, and acquired that mathematical accuracy, that logical preciſion, and ſceptic doubt, which diſtinguiſhed his future life.

From Glaſgow he repaired to Edinburgh, where he was ſoon diſtinguiſhed, and honoured with the friendſhip of Dr. Cullen, who is not more eminently conſpicuous for the ſuperiority of his

 own

own genius, than quick-ſighted in diſtinguiſhing, and liberal in encouraging it in others.

Having finiſhed his ſtudies at Edinburgh, he came to London in the year 1765, and now devoted himſelf entirely to the ſtudy of phyſic; and, looking upon anatomy as one of the principal pillars of the art, he endeavoured to complete with Dr. Hunter, what he had begun with Dr. Monro, and acquired, under this eminent profeſſor, that perfect anatomical knowledge, which appears in all his diſſections. He likewiſe entered himſelf a pupil at St. George's Hoſpital; and, diſguſted, as he has often told me, with the inaccuracy or want of candour, of the generality of practical writers, he determined to obtain an acquaintance with diſeaſes at a better ſchool, and under an abler maſter; and to have, from his own experience, a ſtandard, by which he might judge of the experience of others. With what induſtry he proſecuted this plan, and with what ſucceſs his labours were crowned, we may form ſome judgment from the ſpecimen now offered to the Public.

Whilſt attending the hoſpital, he was alſo employed in making experiments on the blood, and other animal fluids; and likewiſe in a courſe of experiments on chemical pharmacy, which are ſtill extant, and give the ſtrongeſt evidence of his accuracy and diligence; but whether they are of importance enough to be preſented to the public, I have not yet had leiſure to aſcertain.

In the year 1767, he graduated at Leyden, and publiſhed an Inaugural Diſſertation on the Dyſentery. On his return to London he recommenced his ſtudies at the hoſpital; and, in June, 1769, began his

his Experiments on Diet, to which undertaking he was greatly encouraged by Sir John Pringle and Dr. Franklin, whofe friendfhip he then enjoyed, and from whom he received many hints, both as to the plan, and, afterwards, in the execution of his defign. Thefe Experiments, or rather the imprudent zeal with which he profecuted them, proved in the end fatal to himfelf; at leaft, fuch was the general opinion of his friends at the time, but, in my mind, and I fpeak from an intimate knowledge of his character, other caufes, particularly chagrin and difappointment, had no fmall fhare in bringing about this event. Dr. Stark was much more converfant with books than with men; poffeffing great firmnefs and dignity of mind himfelf, with uncommon fimplicity of manners, he was ill prepared for the cold prudence, the time-ferving meannefs, or the bafe duplicity which he met with in others. He had not yet learned in the great fchool of the world, a leffon which all young and ingenuous minds receive at firft with indignation, *viz.* that genius or talents avail nothing, when oppofed to intereft or to faction. Nor had he yet made the obfervation of Figaro, equally applicable to all ages and to all countries,

Que, le favoir faire, vaut mieux que le favoir.

But if Dr. Stark may by fome be reckoned unfortunate, in having been cut off at an early period of life, and before he had obtained that eminence and diftinction to which his talents and application entitled him, he was peculiarly fortunate in what is infinitely more valuable. If his life was fhort, it had, at leaft, been fpent in the moft agreeable, as well as moft ufeful of all purfuits, the purfuit of knowledge. If he did not accumulate wealth, he preferved his independence. If he did not obtain the vain praife of the world, he had the ſuffrage of the wiſe and good, the praiſe that's worth ambition. He enjoyed the high ſatisfaction, laudari a laudatis viris, and a ſtill higher ſatisfaction, in the conſciouſneſs of having always acted his part with integrity and honour; and, in his laſt moments, might have juſtly conſoled himſelf with the magnanimous reflexion of the immortal Tycho, "non inutilis vixi."

For thoſe who wiſh to know his perſon, I ſhall tranſcribe the account he himſelf gives of it, at the beginning of his Experiments on Diet. "The perſon," ſays he, "upon whom theſe Experiments are tried, is a healthy man, about twenty-nine years of age, ſix feet high, ſtoutly made, but not corpulent, of a florid complexion, with red hair."

The character of his mind, which is infinitely more valuable, I ſhall not pretend to delineate; but thoſe who were beſt acquainted with his merit, will not think that I apply improperly to him, what was formerly ſaid by Salluſt, of one of the greateſt and beſt of the Roman citizens—"Non divitiis cum divite, neque factione cum factioſo, ſed cum ſtrenuo virtute, cum modeſto pudore, cum innocente abſtinentia certabat; eſſe, quam videri, bonus malebat*"

* Bel. Catalin. cap. liv.

TABLE OF CONTENTS.

CLINICAL and ANATOMICAL OBSERVATIONS.

PART I.

Diseases of the Stomach, Intestines, and Liver.

CHAP. I.

Diseases of the Stomach illustrated by Dissection.

5 *Stricture*

CHAP. II.

CHAP. III.

PART

PART II.

Diseases of the Chest.

CHAP. I.

Diseases of the Chest illustrated by Dissection.

CHAP.

CHAP. II.

A Description of the Symptoms of Diseases of the Chest, taken from those Cases where the Patients recovered, or where the Author had no Opportunity of examining the Bodies after Death

CHAP. III.

PART III.

Diseases of the Fluids.

CHAP. I.

CHAP. II.

CHAP. III.

PART IV.

Diseases of the Head, Nerves and Muscles.

CHAP. I.

Diseases of the Head, &c. illustrated by Dissection.

CHAP II.

A Description of the Symptoms of Diseases of the Head, Nerves, and Muscles, taken from those Cases where the Patients recovered, or where the Author had no Opportunity of examining the Bodies after Death

CHAP. III.

EXPERIMENTS DIETETICAL AND STATICAL.

EXPERIMENTS ON DIET.

EXPERIMENT I.

EXPERIMENT II.

EXPERIMENT III.

EXPERIMENT IV.

EXPERIMENT V.

EXPERIMENT VI.

EXPERIMENT II. REPEATED.

EXPERIMENT VI. REPEATED.

EXPERIMENT

EXPERIMENT VII.

EXPERIMENT VIII.

EXPERIMENT IX.

EXPERIMENT X.

EXPERIMENT XI.

EXPERIMENT X. REPEATED.

EXPERIMENT XII.

EXPERIMENT XIII.

EXPERIMENT XII. REPEATED.

EXPERIMENT XIV.

EXPERIMENT XV.

EXPERIMENT XVI.

EXPERIMENT XVII.

EXPERIMENT XVIII.

EXPERIMENT

EXPERIMENT VII. REPEATED.

EXPERIMENT XIX.

EXPERIMENT XX.

EXPERIMENT XXI.

EXPERIMENT XXII.

EXPERIMENT XXIII.

EXPERIMENT XXIV.

STATICAL EXPERIMENTS.

CLINICAL

CLINICAL AND ANATOMICAL

OBSERVATIONS.

PREFACE

TO THE

CLINICAL AND ANATOMICAL OBSERVATIONS.

IT is with the greateſt diffidence that the Author of the following work, though encouraged by the advice of ſome very good judges, ventures to preſent it to the Publick, but he flatters himſelf, that in this enlightened age, when original obſervations on diſeaſes are ſo much, and ſo juſtly ſought after, it will be received, at leaſt with indulgence, if not with approbation.

The materials were collected at a large hoſpital, where he had at all times acceſs to the ſick, and, conſequently, the moſt favourable opportunity of obſerving the appearance and progreſs of diſeaſes, and, when they proved fatal, of examining the bodies after death. He employed ſeveral years in theſe reſearches, deeming it an indiſpenſable duty to write a faithful hiſtory of each diſeaſe, from the report of the patient, and never deviating from this rule, but where the ſick were incapable of giving a ſatisfactory account of their complaints; then, and then only, he had recourse to the relation of friends, or of thoſe who were preſent. His remarks are the reſult of obſervation and inſpection, unbiaſſed by any hypotheſis or ſyſtem. He has made but little uſe of the terms of art, from an opinion that they are not always very correctly applied, and being deſirous of avoiding all vain parade of learning, or ground of cavil.

The parts into which this work are divided, are thoſe into which the materials ſeemed naturally to break themſelves, and the order is according to their degree of ſimplicity and certainty.

The firſt place is given to diſeaſes of the alimentary canal; which, as moſt within our reach, are probably the beſt underſtood, and the moſt ſucceſsfully treated.

Next to thoſe of the alimentary canal, which converts our nouriſhment into chyle; are placed the diſeaſes of the heart and lungs, which change the chyle into blood.

The third claſs comprehends the diſeaſes of the blood itſelf, and of the fluids ſecreted from it.

The fourth claſs includes the diſeaſes of the nervous ſyſtem, by far the moſt difficult to be underſtood.

It was the Author's original intention to have added ſeveral other claſſes, particularly one giving an account of common fevers, another on the diſeaſes of the urinary organs, and a third on the diſtempers

diſtempers peculiar to women, but finding it a more difficult and tedious labour than he at firſt apprehended, to abridge diaries of ſingle caſes, and to place thoſe which are ſimilar, in the ſame point of view; he is obliged to defer the execution of this part of his plan to ſome future opportunity.

The firſt chapter contains an account of thoſe diſeaſes which proved fatal, with the morbid appearances upon diſſection.

The ſecond relates the hiſtory of ſymptoms only; for when the diſeaſe did not prove fatal, the morbid ſtate of the parts could not be deſcribed: and, in the preſent imperfect ſtate of the art, however diſcriminating ſymptoms may be, they can only lead to a probable conjecture of the condition of the diſeaſed parts. It would be the perfection, indeed, of medical ſcience, could we, from the ſymptoms alone, declare with certainty, the changes which have taken place in the body, and thus, in many diſorders, have a truly rational foundation for practice. A frequent, careful, and impartial compariſon of the ſymptoms which have preceded death, with the appearances of the dead body, can alone lead to this deſirable perfection.

In the third chapter you have the ſuppoſed effects of medicines. But in this matter, which is of the utmoſt importance, we are liable to great deception, As ſymptoms, whether diſeaſes be left to nature, or treated by art, are always changing, ſometimes favourably, ſometimes unfavourably, ſo that it requires great ſagacity, diligent obſervation, and a thorough knowledge of diſeaſes,

to

to diſtinguiſh between thoſe changes which happen in the natural courſe of a diſtemper, and thoſe which are the effects of remedies applied; yet, unleſs ſuch diſtinction be made, our opinions with regard to the effects of remedies will be perpetually liable to uncertainty. There is, perhaps, no place ſo favourable for obtaining this knowledge as an hoſpital; here we ſee a number of ſick, who, from their circumſtances and ſituation, have not had it in their power to prevent their diſorders from taking their natural courſe; here, therefore, it is, that the foundation muſt be laid of this moſt neceſſary and important diſtinction, whereby numberleſs miſtakes to which this ſubject has always been liable, can alone be obviated.

The examples which I have given of diſeaſes running on in their natural courſe, and terminating favourably, which I call a ſpontaneous cure, will not, I hope, be uſeleſs or uninſtructive.

The delicacy which I am bound to obſerve, when deſcribing the caſes of patients who were under the care, or relating the effects of medicines preſcribed by other phyſicians, obliges me to mention thoſe medicines only which ſeemed ſucceſsful. I am very ſenſible that ſilence, with reſpect to the unſucceſsful caſes, which ſhould be fairly compared with the ſucceſsful, being the proper method of conveying truth and conviction to the mind, is a very great defect, but it was here unavoidable. I have, however, endeavoured to ſupply this deficiency, by relating, and I believe impartially, the reſult of a compariſon made by myſelf. But it will not, I hope, be thought, from my ſilence reſpecting ſeveral remedies whoſe effects appeared ambiguous, that I ſuppoſed them to be wholly inefficacious.

The

The chapters are ſubdivided into ſections, which, in the firſt chapter, conſiſt frequently of one or more caſes, the diſſections not having been ſufficiently numerous to admit of the ſhorter, and more eligible method of compoſing from a number, one general hiſtory.

In the ſecond chapter each article, or ſection, is an attempt towards a general hiſtory of the diſeaſe or ſymptom.

The third chapter is ſometimes divided into ſections, according to the particular remedies, whoſe effects are related.

The queries are thoſe opinions or doubts, which a compariſon of the ſymptoms that preceded, with the appearances after death, ſuggeſted to the author at the time; and were deſigned to direct the attention of the Reader, to the moſt important objects.

The difficulties which attend the execution of ſo extenſive an undertaking, comprehending all the diſorders which come under the care of a phyſician in a large hoſpital, will doubtleſs plead the Author's excuſe with the candid, for the many imperfections of this firſt ſketch, which, at leaſt, has this merit, that it is faithfully copied from nature. Of its numberleſs defects no perſon can be more ſenſible than he is himſelf, but he thinks it better to ſubmit it to its fate, rude and imperfect as it is, than to ſupply any thing from conjecture, that bane of phyſic and bar to all improvement. Upon the whole, he truſts, that this performance, however defective in itſelf, will anſwer one good purpoſe, by pointing out a large hoſpital as an inexhauſtible ſource of the moſt uſeful medical knowledge.

CLINICAL AND ANATOMICAL OBSERVATIONS.

PART I.

Diseases of the Stomach, Intestines, and Liver.

CHAP. I.

Diseases of the Stomach, &c. illustrated by Dissection.

§ 1. *Cancer in the Stomach.*

A MAN, aged forty-five, was seized with a pain about the region of the stomach, attended with purging. At first he voided slime, but afterwards white fibrous substances, in some measure resembling ascarides, together with thin membranes; he complained of want of appetite, low spirits, dimness of sight, and giddiness, which last was so considerable, that he was sometimes in danger of falling down: his pulse was weak and quick, and his strength much impaired, though he continued to walk about till the day of his death, which happened four months after the commencement of his illness. He had then two fainting fits, in the last of which he was carried to bed, and died quietly a few hours afterwards.

On opening the cavity of the abdomen, the ſtomach was found adhering above to the liver, and below to the pancreas. It was diſtended with putrid air, which burſt forth on ſeparating the adheſion at the liver. Great part of its upper ſide was conſumed, and that portion of the liver in contact with the diſeaſed part of the ſtomach, was ulcerated and covered with a putrid reddiſh matter. The lower part of the ſtomach adhering to the pancreas, had ſuffered in like manner. The parts by which it adhered, and ſome others, were thickened, ſoft, ſpongy, and in general rotten. It contained a large firm clot of blood, weighing almoſt a pound, together with ſome putrid matter. The pylorus was greatly thickened, but the paſſage was free. The duodenum and jejunum were of a dark bluiſh colour, but the extremities of the villi of the latter were quite black. The ileum, and large inteſtines, were, to all appearance, ſound.

Was it not ſurprizing that, in the preceding caſe, the patient never complained of ſickneſs, nor was troubled with retchings? Perhaps the ſtomach was too weak for that exertion.

§ 2. *Inflammation of the ſmaller Inteſtines, with Effuſion of Blood.*

A man aged thirty, unknowing of any cauſe, was, one evening, ſuddenly ſeized with retching and vomiting, which were frequent day and night ever after, and conſtantly and immediately followed the ſwallowing any thing, even in the ſmalleſt quantity. His ſkin became yellow on the fourth day, and what he vomited was obſerved on the eighth, to be of a coffee-colour. His ſtools were ſmall, and of a natural appearance, and he had one every day till the tenth. A glyſter being then adminiſtered, a purging with blood enſued. Some of the ſtools conſiſted of clots of blood, with hardly any mixture of other

other ſubſtances: the ſkin and eyes were yellow; his breathing was oppreſſed; the expirations terminated in a ſlight groan, and were about twenty-five in a minute: his tongue, on both ſurfaces, was covered with a whitiſh ſlough; his pulſe was quick and full; he had conſtant ſickneſs, and vaſt uneaſineſs acroſs the ſtomach and hypochondria; the fever encreaſed, the tongue and lips became hard and black, and, retaining his ſenſes to the laſt, he died on the thirteenth day.

On examining the body, the ſtomach and large inteſtines were, externally, of an aſh-colour; the ſmall inteſtines, in general, were of a blackiſh red; towards the lower extremity, of a deep red; and towards the upper extremity, of an olive, or greeniſh brown colour. A portion of the duodenum, a little below the ductus choledochus; alſo a portion of the fundus cœci, were of a very dark red colour, blood being effuſed between the peritoneal and muſcular coats. Internally, The ſtomach, duodenum, and jejunum, were of a dirty brown, or blackiſh colour; the ileum was uniformly of a deep red, and, after being wiped, gave, when preſſed, a reddiſh taint to a white cloth: the rugæ, and ſome other parts of the ſurface of the colon and rectum, were of a light red colour; the parts moſt affected did not, when pulled, appear to be at all weak or tender: the bile was almoſt black; the contents of the ſtomach, and duodenum, were a coffee-coloured fluid: thoſe of the jejunum were a ſoft ſubſtance, like meconium: thoſe of the ileum were a dark red ſubſtance: thoſe of the colon were pure red; the colour of the liver, on its lower ſurface, was dark blue; internally, it was ſound; the gall ducts were quite open: the other parts of the abdomen, and thoſe of the cheſt, were in a ſound ſtate.

§ 3. *Blackneſs of the cellular Subſtance, and Eroſions of the internal and neighbouring Coats of the larger Inteſtines.*

A man, aged fifty, was, when in the Eaſt-Indies, ſeized with a violent bloody flux, accompanied with exceſſive pains in the bowels, and almoſt conſtant ſtraining. The flux, though leſs ſevere, continued almoſt conſtantly for four years afterwards, and he commonly voided very tough ſlime. After that period, it gradually diminiſhed, and had entirely ſtopped about a month before his death; which, as will afterwards appear, (§ 9.) was probably owing to a large abſceſs in the liver.

A woman, aged thirty, after being wet, was ſeized with pains in her limbs and bowels, and, in a week afterwards, with ſo violent a purging, that there was ſometimes not a minute, ſeldom more than ten minutes, and never, ſhe ſaid, above half an hour, between her ſtools: ſhe did not void above a ſpoonful at a time; it was of a natural colour, but frothy and viſcid: the pulſe was very irregular; and the tongue covered with an olive-coloured fur. After a very conſiderable remiſſion, the diſorder returned with the utmoſt violence, and carried her off in about two months from its firſt attack.

In both caſes, the cellular ſubſtance, between the internal and muſcular coats of the large inteſtines, was black, but gave no tinge to water. The blackneſs was either in ſmall circles, or in ſpots, or diffuſed over large portions of the cellular ſubſtance: in the middle of ſuch ſpots as were larger and deeper-coloured than the reſt, the internal coat was

was broken by a very ſmall eroſion. We obſerved other eroſions* a little larger, penetrating into that membrane; they appeared ſometimes white, but moſt commonly black; others, ſtill larger, and which were always white, ran into the muſcular coat: in the parts moſt diſeaſed, more than half the internal coat, great part of the cellular, the tranſverſe and ſome of the longitudinal fibres of the muſcular coat, were conſumed;† externally on the peritoneal coat, faint browniſh ſpots were ſeen oppoſite to the deep black internal ſpots; the ſmall inteſtines were not to appearance diſeaſed.

§ 4. *The glandular Follicles of the great Inteſtines much enlarged, and filled with a glutinous Subſtance.*

A woman, aged twenty-ſeven, was, after an irregular intermittent, ſeized with ſevere purging, accompanied with excruciating pains of the bowels. What ſhe voided, was a thin olive-coloured fluid, with many ſmall portions of a clear glutinous ſubſtance floating in it; they ſomewhat reſembled drops of oil: her pulſe beat commonly about 90 in a minute, and was ſmall; her tongue was uncommonly dry. No conſiderable remiſſion having happened, ſhe died in about ſix weeks after the purging began.

A man, aged fifty-ſix, ſome months after a tedious fever, in which his ſtrength had been greatly impaired, was ſeized with a purging, which, though ſometimes violent, frequently remitted, but never

wholly

* In the inteſtines of the man were obſerved the circles, ſpots, and ſmaller eroſions: in thoſe of the woman, beſides theſe appearances, were obſerved the more advanced changes.

† *See fig.* 1.

wholly ſtopped. Pain of the bowels commonly preceded each fit of purging: what he voided, in the beginning, had been often mixed with blood, but afterwards it was mixed with ſmall maſſes of a clear glutinous ſubſtance, coagulable by heat, or by alcohol, and ſometimes it wholly conſiſted of that ſubſtance. Part of the food, eſpecially liquids, paſſed through the body unaltered: his pulſe beat about 90; his tongue was dry. The diſorder was conſtant and violent for above a month before his death, which happened in eleven months after the purging began.

In the large inteſtines of both, portions of the internal coat were raiſed up into ſmall hemiſpheres, containing a colourleſs glutinous ſubſtance, which was rendered white and firm by alcohol, or by heat, but by cold water was ſoftened, and partly diſſolved.* On the internal ſurface there alſo appeared irregular eminencies and depreſſions, both of which were covered by the internal coat: the former were white, both externally and internally; the latter externally were commonly livid, and ſometimes they were in ſpots: under the eminencies the cellular ſubſtance was thicker and more ſolid; under the depreſſions it was thinner than in the ſound portions of the inteſtines.† In the loweſt part of the ileum, we obſerved eminencies of the ſame colour and ſtructure as thoſe in the colon.‖ There were alſo a few irregular eroſions of the internal coat in the firſt caſe.‡ In the ſecond, we obſerved eroſions ſimilar to thoſe deſcribed under the next article. The other parts of the ſmall inteſtines were in a ſound ſtate.

Were

* *Fig.* 2, 3. *b b*, *c*, *&c.* *d.* iv. & v. *a*, *&c.* *b.*

† *Fig.* 3. *a a.* ‖ *Fig.* 5. *A B.* ‡ *Fig.* 3. *c.*

Were the hemiſpheres, above deſcribed, the inteſtinal glands, enlarged? Was the coagulable part of the blood ſecreted by them, inſtead of common mucus? Is the voiding of a clear gelatinous ſubſtance, in ſmall ſeparate maſſes, the ſign of this ſtate of the inteſtines?

§ 5. *Stricture in the Rectum, and Eroſion of the Glands of the Ileum.*

A man, aged forty-ſix, had almoſt recovered of a flux, which had continued about a year, accompanied with gripings, and after each motion, with blood dropping from him, when, twelve days before he died, the paſſage of his body was entirely ſhut up; the belly ſwelled, and for ſome time was partially puſhed out by portions of the colon. Glyſters and whatever he drank were immediately returned, the former with wind. There was much noiſe in the bowels. The belly becoming at laſt uniformly ſwelled, he died in the utmoſt agony.

The colon was every where diſtended, to almoſt five inches in diameter, by thin fæces and air, which laſt, through ſome ſmall apertures in the coats of the inteſtine, had burſt into the cavity of the abdomen. The ſtomach was compreſſed by a flexure of the colon, which almoſt entirely filled the left hypochondrium, and the diſtenſion of this inteſtine terminated at a ſtricture thereof, a little above the reflexion of the peritoneum over the bladder. At this ſtricturethe paſſage was almoſt wholly ſhut up, by a kind of tubercles, ſoft, ſpongy, and rotten. We obſerved ſome eroſions of the internal coat of the cœcum, and in the lower part of the ileum; alſo eroſions

of

of what is commonly called Peyer's glands;* and, near the attachment of the mesentery, we discovered small holes of the internal coat,† some of which might, by pressing upon the vessels near them, be filled with blood. The other parts of the alimentary canal were internally sound.

§ 6. *Irruption into the Colon of Matter collected near the right Hypochondrium.*

In a man aged thirty-four, during a bloody flux, accompanied with pain chiefly in the upper part of the belly, a hardness was felt, and soon afterwards a tumor appeared near the right hypochondrium. In about three months the flux stopt; but the tumor increased for two months longer, when it broke; and, the opening being enlarged with a knife, discharged about a pint of a thick reddish matter. At this time the purging returned, and in three weeks he voided matter resembling that discharged at the wound, where, while forcing at stool, there was often a bubbling noise. Lying on the left side was soon followed by a motion to stool, and pressure on that side of the belly forced the matter through the anus. The purging increasing, and the discharge, though small in quantity, continuing from the tumor (which had now almost quite subsided), he died in about seven months from the first attack of the purging, and two months after the breaking of the tumor.

The parietes of the abdomen, the edge of the right lobe of the liver, and the neighbouring part of the transverse arch of the colon, were found adhering to one another, and all of them in some degree ulcerated. The colon was at that part perforated by some very

* *Fig.* 7.

† *Fig.* 8.

very ſmall apertures; and its internal coat, in many places, chiefly near the apertures, irregularly eroded.

§ 7. *Numerous Conſtrictions of the Inteſtines.*

A man, aged nineteen, by trade a brazier, having for nine months been often afflicted with pains, ſenſe of twiſting of the bowels, and coſtiveneſs, was ſeized a fortnight before his death with a violent fever, attended in the beginning with purging, and towards the end with ſtupor.

Another man, aged ſixty-five, by trade a houſe-painter, was, for the laſt five years of his life, frequently afflicted with violent pains in his bowels, accompanied with coſtiveneſs: he was oft times nine or ten days without a ſtool. About a month before his death, being greatly weakened and waſted with his diſorder, he was ſeized with a purging, which, though moderate, carried him off.

In both ſubjects, the ſmall, as well as the large inteſtines, were found alternately contracted and dilated: the contracted parts in the former were about one quarter of an inch; in the latter about half an inch in diameter; they were both externally and internally of a deep red colour, ſeemingly from the enlargement of the blood veſſels. The wideſt portions were nearly four times larger than the narroweſt.

§ 8. *Hardneſs of the Liver, and thinneſs of the Bile.*

A man aged thirty, after a fever, became yellow, and continued ſo almoſt conſtantly for four years, his colour being ſenſibly deeper every time he caught cold; but except a ſlight looſeneſs, to which he was now and then ſubject, he had no other remarkable complaint till three weeks before his death. He was then ſeized with thirſt, fever, great pain in the upper part of his belly, and in both hypochondria; a frequent cough, his breathing ſhort and painful, his ſkin and eyes of a deep yellow, his tongue clammy, the greater part of it very red, and one edge of it covered with a white ſlough; his belly was tenſe and ſwollen.

On diſſection the liver appeared large, and of a dirty brown colour: it was hard and uneven on its ſurface, which was raiſed up into ſmall eminencies in cluſters. Externally, it was variegated, dark orange and dark green being mutually interpoſed. On preſſing the gall bladder, a thin tranſparent bile, of a deep green colour, was forced into the duodenum; no ſtone, conſtriction, or other diſeaſed appearance, was found in the gall-ducts or gall-bladder, though both were accurately examined. The omentum was thick and opaque, and turned upwards over the ſtomach and liver: over the ſtomach, it was attached to the diaphragm at its edge; over the liver, by a broad ſurface; but it did not adhere to either of thoſe organs: the ſmall inteſtines adhered by their convolutions to one another, and to the parietes, being covered with a kind of cellular ſubſtance. Internally, they were in a ſound ſtate.

§ 9. *Abſceſs of the Liver.*

A man, who (as already related, § 3.) had been much afflicted with a flux, was, about ſix weeks before his death, ſeized with fits of coldneſs, which came at firſt at irregular periods, but afterwards every forenoon. In the beginning, they laſted four or five hours only, and were followed by heat, head-ach, and thirſt; but towards the end they laſted all day, and were followed by burning heat, continuing throughout the night. The ſtools were of an aſh-colour.

In the right lobe of the liver was an abſceſs, containing about half a pint of matter; the gall-bladder was large, and full of pale yellow bile.

§ 10. *Hydatides in the Liver.*

A man, aged twenty-nine, was, three months before his death, ſeized with pains in the right hypochondrium, ſoon followed by a ſwelling of the part, and yellowneſs of the ſkin, and afterwards by a ſwelling of the whole belly. A fortnight before he died, he was taken with a purging and vomiting, of dark-coloured matter, accompanied with exceſſive gripings. The vomiting ceaſed in a few days, the yellowneſs of the ſkin diſappeared, the belly ſubſided, the pains abated, and the purging only remaining, he, two days before his death, thought himſelf greatly better.

On diſſection, the abdomen contained ſeveral pints of a muddy liquor, tinged yellow; the right lobe of the liver was greatly

 lengthened,

lengthened, its lower ſurface being puſhed out, formed, with the upper one, a continued convexity; and matter iſſued from ſome ſmall apertures on its ſurface. Internally, there were two large cavities, containing about three quarts of a lightiſh brown thick fluid, and many round gelatinous tranſparent bags, white or yellow; the gall-bladder, at the bottom of which was a large opening, was included in the anterior of the two cavities; the gall-ducts were widened, and opened freely into this cavity and into the duodenum. The left lobe of the liver was nearly ſound. A part of the jejunum was reflected over the omentum and colon, adhering to the liver and to the parietes.

§ 11. *Whitiſh Granules, or Tubercles in the Liver.*

On examining the bodies of two perſons, neither of whom had any ſymptom of an affection of the liver, both having laboured under complaints of the cheſt, and which, on diſſection, appeared the chief ſeat of diſeaſe; the liver, though not conſiderably enlarged, throughout its whole extent contained ſmall whitiſh granules, which were not vaſcular but ſmooth, almoſt tranſparent, and ſo numerous, that they occupied more than half the ſpace naturally filled by this viſcus. The gall bladder was very ſmall, and contained little more than a tea-ſpoonful of bile.

§ 12. *The common Gall-duct ſhut up by a Gall-ſtone, and the Hepatic-duct opening into the Duodenum.*

In the body of a man who died of a fever, without having any ſymptom of jaundice, the extremity of the ductus choledochus was quite

quite ſhut up by a large gall-ſtone, which protruded into the duodenum; the gall-bladder was greatly contracted, empty of gall, and covered all over with a cellular ſubſtance; the ductus hepaticus adhered to the duodenum, and opened into it about an inch below the pylorus: over the gall-bladder, the edge of the liver was a little rounded. This viſcus was in other reſpects ſound.

C H A P. II.

A Deſcription of the Symptoms of Diſeaſes of the Stomach, &c. taken from thoſe Caſes where the Patients recovered, or where the Author had no Opportunity of examining the Bodies after Death.

§ 1. *Vomiting.*

BLOOD is ſometimes thrown up by vomiting, mixed with the food, or with the liquors of the ſtomach. The uſual ſymptoms accompanying this, are, giddineſs, pain of the head, and, in ſome inſtances, pain in the left hypochondrium, increaſed, after eating; with exquiſite ſoreneſs, when ſolid food, hot liquors, or hot medicines, are paſſing down into the ſtomach; and, in other caſes, weight at the ſcrobiculus cordis, ſickneſs after eating, till the food is brought up; dimneſs of ſight, diſagreeable dreams, and purging of black matter, or of blood. This diſorder ſometimes follows a blow or a ſprain: moſt commonly the cauſe is unknown. In the firſt inſtance it terminated favourably.

§ 2. *Purging.*

Thoſe caſes of purging which I had an opportunity of obſerving at the hoſpital, were accompanied with thirſt, want of appetite, foulneſs of the tongue, quickneſs of the pulſe, gripings, noiſe in the bowels, ſtraining, with pain in the fundament; and ſometimes the food paſſed unaltered. They may, according to the matter voided, be divided into two ſpecies; the ſlimy, and the gelatinous: in both of which blood is frequently paſſed. In the ſlimy purging, the ſtools are frothy, and conſiſt of a yellowiſh or whitiſh, viſcid, ropy matter; which, unmixed, is ſometimes, with the utmoſt ſtraining, forced off ſcalding hot, and ſometimes paſſes off almoſt involuntarily. This ſpecies is often accompanied with darkneſs before the eyes, giddineſs, retching, vomiting. In the gelatinous purging, the ſtools are either a thin liquid, containing ſmall, clear, whitiſh gelatinous ſubſtances, or almoſt wholly jelly; and ſometimes nothing but wind is voided. In this ſpecies, any liquid taken into the body is apt immediately to run off with ſevere gripings. The duration of either ſpecies is, moſt commonly, a few weeks; ſometimes two or three months: and, in one caſe, the ſlimy purging continued three years. The event is frequently fatal. The cauſes are uncertain.

Is it not probable that, in the firſt ſpecies, the bowels are in the ſtate deſcribed (§. 3.); that, in the ſecond, they are in the ſtate deſcribed (§. 4.); and that, when blood is voided, which happens in either ſpecies, it proceeds from the eroſion of blood-veſſels, as deſcribed (§. 5.)?

§ 3. *Coſtiveneſs.*

The almoſt conſtant attendants on coſtiveneſs, are head-ach, ſickneſs, vomiting. It is accompanied alſo with ſlight pricking, or ſevere and violent pains; either in the right ſide of the belly, near the ſcrobiculus cordis; in the left ſide of the belly, near the anus; or over the whole belly: theſe pains are oft times increaſed by ſlight preſſure of the part. When the diſorder is moſt violent, it ſometimes has exacerbations, in which the bowels, after a ſenſe of coldneſs in them, are, according to the feelings of the patient, twiſted, drawn together, and ſqueezed to the back; the teeth gnaſh, the body is drawn forward; whatever is then taken is immediately vomited, and glyſters adminiſtered are returned without fæces. As the ſtools are, for the moſt part, retained till the remedies given have proved effectual, the coſtiveneſs, in ſome violent caſes, has continued a fortnight or a month; and one patient, a painter, had no ſtool for three months. In general, purging medicines, and glyſters, when retained, produce very ſoon the deſired effect. The duration of the diſeaſe is various and uncertain, but the moſt obſtinate caſes continue ſometimes two or three years; intervals, which now and then happen after a ſpontaneous purging, being interpoſed. The cauſe, even when the diſorder is moſt violent, is, at times, altogether unknown. Lead, in various forms, and the fumes of quickſilver, frequently occaſion it in painters, and other workmen, who, in their ſeveral trades, make uſe of thoſe metals. It ſometimes follows a blow on the belly, or a ſprain in lifting a great weight; and, in this caſe, blood, or matter, is voided with hardened fæces. It did not, in any of the inſtances from whence this hiſtory is drawn, prove fatal.

Are the bowels, in ſuch caſes, ſometimes inflamed? Were they, in the more violent caſes, nearly in the ſtate deſcribed (§. 7.)?

§ 4. *Jaundice.*

This diſeaſe begins with ſickneſs and pain at the ſcrobiculus cordis, and ſometimes with giddineſs, retching, and vomiting of a yellow ſour ropy phlegm. The urine is of a ſaffron colour, ſtaining paper that is dipt in it, and becomes turbid when cold. The tunica ſclerotica of the eye and ſkin are yellow; the yellowneſs is firſt perceptible at the pit of the ſtomach; external objects appear as uſual. This diſeaſe is frequently accompanied with purging, though ſometimes with coſtiveneſs: the ſtools are commonly in colour like blue clay; ſometimes of a dark earthy, or of a deep yellow colour; but, ſo far as I have ſeen, never white. In general, there is a bad taſte in the mouth, with a white tongue and a ſlight fever; frequently, immediately preceding the vomiting, there is a violent pain in the back; the pains in the belly are increaſed by walking; they complain of pain in the right, or left hypochondrium, or in the flank; and ſometimes, though rarely, of pains ſhooting from the ſhoulder to the breaſt, or from the back down the thighs. In ſome caſes, the diſeaſe intermits for ſeveral weeks or months, but more commonly is continued, though the pains and vomiting attack by fits, laſting either a few hours every morning, or for ſeveral days. Relief always follows ſpontaneous vomiting, or purging. When the diſeaſe is going off, there is ſometimes a violent itching of the ſkin.

Does not the relief which follows ſpontaneous vomiting and purging, point out the proper method of curing this diſorder, by emetics and purgatives?

C H A P. III.

Obſervations on the Effect of Remedies, given in the Cure of Diſeaſes of the Stomach, &c.

OPIUM alone ſeldom failed to reſtrain purgings for two or three days or a week; but the diſeaſe, at the end of thoſe periods, returned, and commonly with more violence, than before opium had been taken. But, though this drug alone appeared to be a medicine altogether inadequate to the cure of purging, yet when combined with others, moſt excellent medicines were formed, whoſe effects were not leſs powerful, and were more laſting.

In the ſlimy purging, the moſt efficacious medicine was vitri antimonii cerati* gr. v. opii circiter gr. i. quotidie. Another powerful medicine was radicis columbæ gr. x. opii gr. i. in die. Columba root alone gave only a temporary relief. In the gelatinous purging the moſt efficacious medicine was ipec. gr. i. opii gr. i. quotidie. Vitr. antim. cerat. cum opio, given in this ſpecies, aggravated the ſymptoms. Other uſeful medicines, in either ſpecies, but of inferior efficacy, were opium with rhubarb, with aromatics, with abſorbents, or the abſorbents alone. When the pains were violent, fomentations gave much relief. In coſtiveneſs, the beſt remedies were fomentations and the common purgatives. In the painter's colick, oil, or oil with rhubarb, was moſt uſeful. In the jaundice, emetics and purgatives were uſeful remedies.

* A medicine in the Edinburgh Pharmacopœia.

Is not the combination of opium with other drugs, recommended to us by practitioners in all ages, and of all sects? Have we not an example of this in the antient compositions Mithridate, Theriaca, and several others, which are still retained in the modern dispensatories, and in which opium is a principal ingredient; also, in the highly celebrated medicines of Dover, and of Ward, the most efficacious of which are opium joined with ipecacuan, with hellebore, or with mercury.

PART II.

PART II.

Diseases of the Chest.

CHAP. I.

Diseases of the Chest, illustrated by Dissection.

§ 1. *The Canal of the Aorta almost shut up by the semilunar Valves.*

A WOMAN, aged twenty-one, who got her bread by hard labour, had, for five years, been subject to fits of palpitation, which attacked her commonly after an interval of some months: the last fit, in which she died, lasted five weeks, being more violent and of longer continuance than any of the preceding ones. In this fit, the left hypochondrium and scrobiculus cordis were much pushed out at each palpitation; there was also a remarkable throbbing in the course of the vessels on each side of the neck, but from the irregularity of those motions, they could not be counted, and the parts themselves were so tender, that she would hardly allow them to be touched. Her pulse was weak, quick, and irregular, sometimes fluttering, sometimes intermitting: she complained of pain and tightness across the chest; her breathing was oppressed and quick, inspiring commonly forty-five times in a minute; she had a short cough, was low, faint, constantly sick, and, for most part, vomited immediately after swallowing the smallest quantity of any thing, whether liquid or solid. At first, she lay on her left side, or on her back; afterwards on her back only, having her head and

 shoulders

ſhoulders raiſed up, and at laſt with her arms folded over her head. She became anaſarcous a month, and yellow two or three days before her death.

On diſſection, the lungs were found adhering to the pericardium, and to part of the parietes of the cheſt near it: in other places detached, every where ſoft, and, when preſſed, froth iſſued out of the noſtrils. The ſemilunar valves of the heart were thickened, and projected towards the axis of the aorta.* The heart was ſeemingly lengthened, in other reſpects ſound. The large blood-veſſels, which were traced and cut up, as far as the head and arm-pits, were alſo ſound. The abdominal viſcera were in a natural ſtate.

§ 2. *Pericardium adhering to the Heart, &c.*

A woman, aged twenty-ſeven, was, ſome months before death, ſeized with a frequent dry cough, followed by pain in the left hypochondrium, and at the ſcrobiculus cordis: her breathing was ſhort and quick, her pulſe commonly one hundred in a minute; ſhe complained of ſickneſs, with conſtant and often violent head-ach. A fortnight before her death, ſhe loſt the uſe, firſt, of the left arm; then of all the left ſide, and her ſpeech faultered.

The pericardium was found adhering every where to the heart, which was much enlarged, and hardened, but internally ſound; the lower, and greater part of the lungs of the left ſide, were of a dark red colour, firm, and adhered to the neighbouring parts; there was a very ſmall quantity of water in the right cavity of the cheſt. The ſtomach

* *Fig.* 9, 10.

ſtomach was narrow: the other abdominal viſcera had a natural appearance.

§ 3. *The Pericardium enlarged, containing eight Ounces of a Fluid, and, by fatty Papillæ, adhering partially to the Heart.*

A girl, aged fourteen, was, three weeks before her death, ſeized with great difficulty of breathing, and with pain in the left ſide, attended ſometimes with a ſhort cough, which was not at all relieved by repeated venæſection. When in bed ſhe lay conſtantly on the left ſide, her cough being excited by any attempt to lye on the right ſide, or on her back. She often choſe to ſit up; but whether ſitting or lying, the body was always much bent forward. The pulſe was full, and very quick.

The pericardium was much enlarged, and being covered towards the upper part with a ſoft ſubſtance half an inch thick, concealed all the lungs of the left ſide, except a ſmall portion of the upper lobe near its edge. It contained eight ounces of a fluid: its internal ſurface, and alſo the external ſurface of the heart, was covered, in many places, with a layer of a kind of fatty matter, eaſily ſeparable from either ſurface, and ſupporting numerous oblong fatty papillæ. The oppoſite papillæ on the pericardium, and on the heart, in ſome places, adhered to one another. The great veſſels within the pericardium, were covered by a ſoft ſubſtance a quarter of an inch thick. The heart and great veſſels, internally, were ſound. The lungs adhered univerſally, though ſlightly; were in every part ſoft, and eaſily dilated by air blown in by the windpipe. There were ſome ounces of a fluid in

each

each cavity of the cheſt. The abdominal viſcera, except the right kidney, were quite ſound.

On examining the body of another woman who died of a conſumption, but who, a fortnight before her death, had lain alſo, night and day, bent forward on her elbows and knees, the pericardium contained much water.

§ 4. *Ulceration of the Lungs, or pulmonary Conſumption.*

The frequency and fatality of this diſorder having afforded me many opportunities of obſerving the ſymptoms, and of examining the ſtate of the body after death, I ſhall here, inſtead of particular inſtances, endeavour to give a general deſcription of the ſymptoms, and of the appearances on diſſection, taken from ten caſes, where the diſeaſe proved fatal.

Symptoms of the Diſeaſe.

The ſymptoms of the diſeaſe may be divided into primary and ſecondary; the former being ſuch as are peculiar to affections of the cheſt, the latter, ſuch as are common to thoſe, and to ſome other affections.

Of the firſt kind are cough, ſpitting, pains of the cheſt, difficult breathing, and poſture. Of the ſecond kind may be reckoned coldneſs, heat, ſweating, purging, waſting, pains of the limbs, &c.

The cough, which is brought on by expoſure to cold, or by drinking any cold liquor whilſt hot, or by various other cauſes, is almoſt

almoſt conſtantly the firſt ſymptom, and in the beginning often the only one; though it is, at times, accompanied with ſtitches, or ſhooting pains in the cheſt, and with expectoration. It generally attacks by fits, which are moſt frequent and ſevere towards evening, or during the night, preventing ſleep.

The ſpitting or expectoration, is commonly very thick and viſcid, of an aſh-colour, with a ſlight tinge of green, and contains many air bubbles; ſometimes it is yellowiſh, and in ſmall round maſſes, which probably come from ſmall vomicæ; now-and-then, though rarely, it is ſtreaked with blood. The quantity expectorated is generally inconſiderable in the beginning, but afterwards increaſes to about half a pint, or a pint, in twenty-four hours. In thoſe caſes, where (upon diſſection) the large vomicæ were found almoſt empty, the ſpitting, towards the end, had been in very ſmall quantity.

As the ſpitting is, perhaps, the moſt certain criterion of vomica, it will be proper to enquire into its peculiar character, that it may be diſtinguiſhed from pus and mucus: two ſubſtances which it greatly reſembles. All of them, when free from air bubbles, ſink in water. Pus is eaſily diffuſible in it, by gentle agitation, but in a few hours falls to the bottom. Mucus cannot be equally diffuſed in water without ſtrong agitation, but when diffuſed, forms with it a permanent ropy liquor. The ſpitting of conſumptive perſons is diffuſible in water more eaſily than mucus, and like that, at firſt forms with it a permanent ropy liquor; but which, in a few days, depoſits a ſediment in the ſame manner as pus; the liquor, however, ſtill continuing ropy, and reſembling mucus and water.

The pains of the cheſt are of two ſorts; viz. ſtitches, which ſometimes come on in the beginning; or a general ſoreneſs

of

of the cheſt, which is moſt ſeverely felt after violent fits of coughing.

The breathing (even before the diſeaſe has arrived at its acme) is generally two or three times more frequent than that of a perſon in health, and is often accompanied with a ſighing noiſe, and performed with great motion of the cheſt; but it is ſomewhat relieved by the expectoration which follows the fits of coughing. Neither inſpiration, nor expiration, can be continued ſo long as by a healthy perſon; but the former, in conſequence of the pain or cough excited by it, is moſt ſenſibly ſhortened.

With reſpect to poſture, the patient commonly lies on his right ſide; but this is not compleatly fixed till the diſeaſe is far advanced, when he can only lie on his back, with his head and ſhoulders high, and ſometimes with his knees drawn up.

The coldneſs (which ſometimes precedes any ſigns of an affection of the cheſt) comes on by fits, either regularly every day, or every other day, like the paroxyſms of an intermittent fever; or, as is moſt common, at uncertain periods.

The heat is of two kinds, either a burning heat, with intenſe thirſt, continuing all night, which ſucceeds the fits of coldneſs; or a continued heat, increaſing towards evening, which, in general, is much more moderate.

The pulſe is always ſmall and quick; commonly there is a loſs of appetite, though, in ſome inſtances, towards the end of the diſorder, the appetite is voracious.

The

The ſweating is almoſt a conſtant ſymptom, and is at times profuſe, breaking forth, chiefly, on the head and breaſt; though more commonly it is moderate, and follows the evening exacerbation; and ſometimes towards the end, it diminiſhes, or ceaſes.

The purging ſeldom comes on till near the end of the diſeaſe, at which time the legs are apt to ſwell. When the purging begins all the feveriſh ſymptoms greatly abate, but are again increaſed, if, by any means, it is ſtopped.

The waſting of the body is more remarkable in this, than in any other diſeaſe.

Pains in the limbs, or all over the body, are alſo not unfrequent ſymptoms; and the menſes, in women, (who are more liable to this diſeaſe than men) commonly ceaſe ſoon after it is eſtabliſhed.

The duration of the diſeaſe is various, from four months to two years; and it will be found to be nearly in proportion to the age of the patients, which varies from ſeventeen to thirty-five years.

Appearances on Diſſection.

As the appearances on diſſection, though extremely uniform, are very different in degree, it may be uſeful to arrange them under the following heads:—Tubercle; Vomica; State of the Air Veſicles, and cellular Subſtance; State of the large Blood Veſſels; Trachea; the Degrees of morbid Affection; and ſome other circumſtances.—

Tubercle.

In the cellular ſubſtance of the lungs are found roundiſh firm bodies, (named tubercles) of different ſizes, from the ſmalleſt granule, to about half an inch in diameter; the latter often in cluſters. The tubercles of a ſmall ſize are always ſolid, even thoſe of a larger are frequently ſo; they are of a whitiſh colour, and of a conſiſtence approaching nearly to the hardneſs of cartilage; when cut through, the ſurface appears ſmooth, ſhining, and uniform. No veſicles, cells, or veſſels are to be ſeen in them, even when examined with a microſcope, after injecting the pulmonary artery and vein. On the cut ſurface of ſome tubercles were obſerved ſmall holes, as if made by the pricking of a pin; in others were found one or more ſmall cavities, containing a thick white fluid, like pus; at the bottom alſo of each of theſe cavities, when emptied, ſeveral ſmall holes were frequently to be ſeen, from which, on preſſing the tubercle, matter iſſued; but neither theſe holes, nor the others abovementioned, (ſo far at leaſt as could be determined) communicated with any veſſels. The cavities, in different tubercles, are of different ſizes, from the ſmalleſt perceptible, to half an inch, or three quarters of an inch, in diameter; and, when cut through and emptied, have the appearance of ſmall white cups, nothing remaining of the ſubſtance of the tubercle, except a thin covering or capſula. The cavities of leſs than half an inch diameter are always quite ſhut up; thoſe which are a little larger have, as conſtantly, a round opening made by a branch of the trachea. At this period, there being a free paſſage for the matter contained in the tubercle into the trachea, and a communication between the cavity of it and the open air, it is proper to change the name of tubercle to that of vomica.

Vomica.

Vomica.

The ſmaller vomicæ are commonly entire, the larger are frequently ruptured; the largeſt (which, generally ſpeaking, are of an oval ſhape, and about four inches in length) are lined, either partially, or entirely, with a ſmooth, thin, tender ſlough or membrane; the ſame as the capſula of the ſmaller vomicæ. The matter contained in them, when the capſula is entire, is whitiſh or yellowiſh; when ruptured, reddiſh; in either caſe readily diffuſible in water. It is proper, however, to remark, that even in the largeſt vomicæ, when they are not compleatly ruptured, the matter is ſeldom red, but yellowiſh, aſh-coloured, or greeniſh; often fœtid.

Into all vomicæ (the ſmalleſt perhaps excepted) there are ſeveral openings of the bronchia; alſo openings forming communications between the different vomicæ; the bronchial openings are commonly round and ſmooth; the others, generally irregular and ragged. The larger vomicæ, which have numerous bronchial openings, are found to contain ſcarcely more matter than is ſufficient to beſmear their ſurface; and what ſhews clearly that the matter of vomicæ is diſcharged by theſe openings of the aſpera arteria, is, that if a deep inciſion be made into any diſeaſed part of the lungs, and that part gently compreſſed, the matter will be ſeen to iſſue from the cut extremities of the bronchia; or if any conſiderable branch of the aſpera arteria be laid open, and the lungs preſſed in the ſame manner, the matter will be ſeen coming into it, from the ſmaller ramifications.

The largeſt vomicæ are generally ſituated towards the back part of either upper lobe, and are commonly concealed; though ſometimes on the ſurface of that part of the lungs, which is thin and ſinks into a hollow, there are ſeveral ſmall apertures leading to the vomica; and ſometimes, though rarely, a vomica is a hemiſpherical cavity on the outward part of the lungs. Wherever there is a vomica there is always a broad and firm adheſion of that part of the lungs to the parietes, or pleura, ſo as to preclude all communication between the cavity of the vomica and that of the cheſt; even tubercles are ſeldom found without adheſion.

State of the Air Veſicles, and cellular Subſtance.

Thoſe parts of the lungs which are contiguous to tubercles are red, ſometimes ſoft, but more frequently firm or hard; and whilſt other parts of the lungs unaffected by diſeaſe are readily diſtended, by blowing into the trachea, thoſe portions which are contiguous to tubercles or vomicæ, remain depreſſed and impervious to air, either blown into the lungs in this manner, or forced, by a blow-pipe, into inciſions made on the ſurface. So that the function of the lungs, ſo far as reſpects the admiſſion of air, ſeems, in thoſe parts, entirely deſtroyed.

State of the large Blood Veſſels.

The pulmonary arteries and veins, as they approach the larger vomicæ are ſuddenly contracted; a blood veſſel, which, at its beginning, meaſured nearly half an inch in circumference, ſometimes (though it had ſent off no conſiderable branch) could not be cut up farther than an inch; and when, outwardly, they are of a larger ſize, yet, internally, they have a very ſmall canal, being almoſt filled up by a fibrous ſubſtance; and frequently, as they paſs along the ſides of vomicæ, they are found quite detached, for about an inch of their courſe,

courſe, from the neighbouring parts. That the blood veſſels are thus obſtructed, and that they have little or no communication with the vomicæ, is rendered ſtill more evident, by blowing into them, or injecting them; by blowing they are not ſenſibly diſtended, nor does the air paſs into the vomicæ, excepting very rarely, and then only by ſome imperceptible holes; and, after injecting the lungs by the pulmonary artery and vein, the parts, leſs affected by diſeaſe, which before injection were the ſofteſt, become the hardeſt; and, *vice verſa*, the moſt diſeaſed parts, before injection the hardeſt, are now the ſofteſt. Upon cutting into the ſounder parts, numberleſs ramuli may be ſeen, filled with the wax, but in the diſeaſed parts there is no ſuch appearance; and upon tracing, by diſſection, the injected veſſels, thoſe which terminate in the ſounder parts may be traced for a long way to the ſmaller ramuli, but thoſe which lead to tubercles and vomicæ, a very ſhort way, and only to their principal branches. The wax was very rarely found to have entered the middling ſized vomicæ, and never the ſmaller or larger ones.

Trachea.

The branches of the trachea are never found in any degree contracted; the internal ſurface of thoſe which opened into the large vomicæ, was of a deep red, (ſeemingly from the enlargement of veſſels) and the internal ſurface of the trachea itſelf, was ſometimes partially red.

The Degrees of morbid Affection.

The degrees of morbid affection are very different, in different ſubjects, and in different parts of the lungs of the ſame ſubject. In ſome caſes cafes there are no vomicæ to be found above an inch in diameter; in others, feveral of two, three or four inches. In the former cafes, the pulmonary arteries and veins are hardly fenfibly contracted. Sometimes not above a third or fourth part of the lungs are affected; at other times, the lungs, of one or both fides, are entirely difeafed. From a rude calculation made on difeafed lungs, the part which remained fit for the admiffion of air, may be eftimated, at a medium, to be about one fourth of the whole fubftance of the lungs. When the lungs are only partially affected by difeafe, the difeafed parts are always the higher, and rather the pofterior; whilft the found parts are the lower, and rather the anterior. When they are wholly difeafed, the higher and pofterior parts, are always much more fo than the reft; and the lungs of the left fide are more commonly affected than thofe of the right.

The lymphatic glands in the cheft are frequently blackifh, and fometimes contain a fubftance like moiftened chalk. In the abdomen there is not any thing remarkable, excepting, fometimes, flight erofions of the villous coat of the inteftines.

Is a conftant cough, though unaccompanied with any other complaint, a fymptom of tubercles in the lungs? Is it, when attended with fits of coldnefs, and with fpitting, a certain fign of vomicæ? Is not the fpitting compofed of matter from the vomicæ, and of mucus from the membrane of the trachea? Does not the contracted ftate of the pulmonary veffels, and the thickening of their coats, prevent, in moft cafes, the fatal hæmorrhages, which otherwife would enfue? Is there not fome reafon to apprehend, that though a tranfitory relief is fometimes afforded by fmall bleedings, the progrefs of the difeafe is thereby quickened?

§ 5. *An*

§ 5. *An Ancurism of the pulmonary Artery opening into a Vomica.*

A man, aged twenty-nine, who had led a very irregular and riotous life, was, for ten months before his death, ſubject to a ſlight cough, which came on immediately after his recovery from the meaſles. Notwithſtanding his cough, he purſued his uſual courſe of life; and, three weeks before his death, was taken ill in the night, with a violent bleeding at the mouth and noſe, which continued about a quarter of an hour, and returned four times at different intervals. He was pale, weak, faint, low-ſpirited, and apprehenſive of death, but breathed eaſily and coughed ſeldom. The night before his death he reſted well, and roſe in the morning without any particular complaint; but, having again laid down in bed, he was, when aſleep, ſeized with a fit of coughing, and blood began to flow (interruptedly), but without any effort from his mouth, though, ſometimes, it was brought up by a ſlight cough, or blown haſtily from his noſe. When the bleeding began, he immediately got up, and ſat upon the bed, although he could not continue for a moment in the ſame poſture, but was conſtantly either bending forwards, or reclining from ſide to ſide. At laſt, in a profuſe ſweat, he ſtarted upon his legs, and, with amazing quickneſs, threw off his waiſtcoat: the cough and bleeding immediately ceaſed; his pulſe, which before had been very quick, was not now to be felt; his thighs trembled, his urine ran from him, and he ſunk down into the arms of a perſon who was ſtanding bye, dying without a ſigh or a groan, in about ten minutes from the time the hæmorrhage began: the quantity of blood which he loſt, was about a quart.

In

In a branch of the left pulmonary artery, which paſſed along a vomica, in the upper and poſterior part of the lungs of the left ſide, was an aneuriſmal ſac, about an inch long, and one third of an inch broad: the coats of the ſac reſembled thoſe of the artery, only thicker; on one ſide of it was a ſlit, with coagulated blood adhering to it, both internally and externally; within the ſac, the coagulum was ſomewhat whitiſh: externally, it was divided into three branches, formed by three ramifications of the aſpera arteria, that opened into the vomica; the other ramifications of the aſpera arteria, and even the trunk itſelf, being alſo filled with coagulated blood. There was no blood in the veſicles of the lungs, which were, every where, evidently diſtended with air, and the air, upon preſſure, readily paſſed from one lobule to another, but could not be forced out at any branch of the trachea, except at the vomica above-mentioned. On opening the cheſt, the lungs did not ſubſide; they were of a light grey colour, with many ſmall aſh-coloured granules, but no adheſion of their ſurface, no other vomica, tubercle, or hardneſs, in any part of them. There was no blood in any of the cavities of the heart, excepting a few ſmall clots between the carneæ columnæ. In the large blood veſſels, which iſſue immediately from the heart, there were ſome very ſmall polypi. The ſubclavian vein was empty; the abdominal viſcera ſound.

§ 6. *The Veſicles of the Lungs filled with extravaſed Blood.*

Three middle-aged men were, all of them, ſeized, ſome months before they died, with pains in the cheſt, which, in two of them, were ſevere from the beginning: in the third, moderate till within three weeks of his death. They were accompanied with ſhivering and vomiting; the ſhivering recurred at intervals, commonly every morning, and was followed by head-ach, heat, and profuſe ſweating; the

the patients commonly lay, and with moſt eaſe, on the ſide principally affected, excepting in the night, when they were ſometimes obliged to ſit up. The breathing was about twice as quick as uſual, and the expirations ended with a ſlight groan. The cough was very frequent, and in one caſe almoſt conſtant. The fever was high in all, and in two inſtances attended with delirium. The pulſe, from ninety to one hundred and twenty in a minute, was full; and, at laſt, beat with a kind of vibration. Pure blood burſt forth, or was brought up in conſiderable quantity by coughing: in one patient, about three weeks; in another, about one week; and in the third (who alone had been repeatedly blooded in the beginning) only two days before his death. Two of thoſe men, one of whom had lived rather faſt, and was ſubject to a cough in winter, became anaſarcous ſome weeks before the fatal concluſion of their illneſs.

The air veſicles, in ſome parts of the lungs, were filled with blood, or with bloody ſerum: thoſe parts did not collapſe on opening the thorax; they were firm, and of a very dark, or of a light red colour; they could not be compreſſed, nor was it poſſible to diſtend them with air blown in from the windpipe, or at punctures made on the ſurface. In ſome inſtances, however, they did collapſe, and admitted to a certain degree of compreſſion or diſtention. The lungs themſelves were ſurrounded by a bloody fluid, the quantity of which varied from a few ounces to ſeveral pints; they were frequently attached to the ſides by membranous adheſions; when cut into, a thick blood, or bloody matter, iſſued forth at the cut ſurfaces; and ſlices cut off from the diſeaſed parts, after having for ſome time been macerated in water, ſtill ſunk in it, in the ſame manner as before maceration. The inſide of the trachea was pale red.

The parts of the lungs chiefly affected, in the preceding caſes, were, in one caſe, the whole lungs of the left ſide, beſides a large

 quantity

quantity of fluid in the cavity: in another, the poſterior part of the upper and middle lobes of the left ſide; alſo the whole of the lower lobe of the right: in a third, the whole of the lungs in both ſides were diſeaſed, although thoſe in the right were moſt conſiderably ſo; in this caſe only, the bloody matter, as mentioned above, iſſued at the inciſions made in the lungs. There were no other præternatural appearances, excepting in one body, where the liver was hard and granulated.

To aſcertain more accurately the ſtate of the air and blood-veſſels, the following trials were repeatedly made on two portions of the lungs taken from the ſame body; one of which was apparently ſound, the other ſlightly diſeaſed. On the cut ſurface of each portion, air was forced in by a blow-pipe; through the largeſt branch, we could find of the pulmonary artery, vein, and aſpera arteria. Upon blowing into the branch of the pulmonary artery, in the diſeaſed portion, the minuter ramuli were diſtended, and a little air bubbled out at ſome very minute openings on the cut ſurface. Upon blowing into the branch of the pulmonary vein, the air veſicles were diſtended, and air bubbled forth at the largeſt bronchial orifice; and, upon blowing into this laſt, the air veſicles were diſtended, and air eſcaped, with ſome blood, at the large venal branch. The ſame experiments being made on the ſound portion of the lungs, the event was ſomewhat different; for, upon blowing into the arterial, or venal branch, the ramuli peculiar to each were alone diſtended, and a little air eſcaped at ſome minute openings on the cut ſurface. Upon blowing into the branch of the trachea, the air veſicles were diſtended, and no air eſcaped. The ſame experiment was alſo repeated on the ſound lungs of another ſubject, and with the ſame effect.

§ 7. *Lymph in the Thorax.*

Three men, two of them middle-aged, the third ſixty-five, were afflicted with a cough, attended with a frothy expectoration: two of them had this complaint for ſome months; the third, who had lived rather irregularly for ſome years before his death. They were out of breath upon walking only a few yards, and ſpeaking was ſo trouble-ſome to them, that they were unwilling to give any account of their feelings; their breathing was quick, and the expirations ſometimes terminated in a ſlight groan; they could blow but feebly, and for a ſhort time: they were, in general, deſirous to ſit up; and, when prevailed on to lie in bed, they were reſtleſs; or, if they continued for any time in one poſture, it was lying on the back with the head high, or on the ſide in which (as afterwards appeared) the fluid was contained. The pulſe was very quick and ſmall; two of them had an inconſiderable ſwelling of the belly and ancles: and thoſe two who had the lungs hardened, were hoarſe; the other was not.

A yellowiſh tranſparent fluid was found in one or in both cavities of the cheſt; it coagulated by heat, though leſs firmly than the ſerum, having a larger proportion of water; the quantity of this fluid, in each cavity, was nearly a pint; the lungs were, more or leſs, diſeaſed in all, with partial adheſions of the higher parts of them to the parietes: in one caſe, there were only ſome ſmall tubercles in the higher part of the upper lobe; in the other two caſes, the whole of the upper lobes, and part of the lower, were very hard, could not be diſtended by air, and when cut into, emitted a bloody froth. In one caſe, we obſerved on the ſurface of the lungs, ſmall bliſters,

 containing

containing a clear fluid. In two bodies, there was a ſmall quantity of water in the abdomen; and, in one, the liver was granulated, the omentum in folds.

§ 8. *Inflammation of the Pleura, and Effuſion of Blood in the intercoſtal Muſcles.*

A woman, aged thirty, who for three months had been afflicted with ſevere purging, had alſo, ſoon after this complaint began, been taken with a cough, at firſt accompanied with ſpitting of blood, but afterwards of thick mucus and purulent matter. About a month before her death, when greatly weakened by theſe complaints, ſhe was ſeized with violent pains, or ſtitches, in the left ſide, which almoſt entirely prevented her breathing: her pulſe, as before, was quick, ſmall, and weak: two bliſters having been applied, the pains, in ſix days, abated, and afterwards were only felt on coughing; during the violence of the complaint, ſhe lay on the ſide affected, but towards the end, her breathing being very ſhort and difficult, eſpecially in the night, ſhe ſat bolſtered up in bed.

In the left ſide of the thorax, the lungs were of a very dark red colour, particularly at the upper part, where we found a vomica, and ſome tubercles: there were alſo ſome adheſions at this part, and at this part only; the pleura lining the ribs, was ſmooth, but its poſterior part, particularly where contiguous to the intercoſtal muſcles, was of a dark red; the redneſs penetrated the muſcles, and, in ſome places, extended to the ſerratus major; it ſeemed partly owing to the enlargement of blood-veſſels, but principally to an effuſion of blood into the cellular ſubſtance, and which, by preſſure, could be forced from one part to another. In the right ſide, the lungs,

excepting

excepting a few tubercles in their upper part, were ſound, and free from adheſion, nor was there any redneſs of the pleura. In each cavity there was about a pint of yellow ſerum, though the quantity was greater in the right than in the left. The inteſtines adhered externally to one another, and there was a ſlight redneſs to be ſeen on ſome parts of their internal ſurface.

§ 9. *Suppuration of the contiguous Surfaces of the Diaphragm and Liver.*

A blackſmith, aged fifty, having, in the depth of winter, lain ſeveral nights in a cold houſe upon ſtraw, was, two months before his death, ſeized with pains acroſs the lower part of the cheſt, difficulty in breathing, and cough, but without ſpitting. The pains fixed in the right hypochondrium, and were, ſometimes, felt at the ſcrobiculus cordis. In ſpeaking, he could only whiſper, but was not hoarſe. The cough was performed with very little noiſe, and reſembled more a lengthened-out expiration than common coughing. He could ſuck in air, or blow it through a quill a long time, and without pain: his pulſe was low; he lay on either ſide, or on his back, and often with the body bent forward, his chin reſting upon his breaſt. Sometimes he was obliged to ſit up, eſpecially a few days before his death, when he could not utter above two or three words without ſtopping, and ſaid, he could hardly breathe, but had then no pain.

In the right hypochondrium, the greater part of the contiguous ſurfaces of the diaphragm and liver, were inflamed and covered with purulent matter; but the inflammation did not penetrate into the ſubſtance of either organ; there was no other preternatural appearance in the abdomen;

abdomen; the lungs, and other parts of the cheſt, were accurately examined, and ſound to be in every reſpect ſound, excepting a few ſlight adheſions in the right ſide.

CHAP. II.

A Deſcription of the Symptoms of Diſeaſes of the Cheſt: taken from thoſe Caſes where the Patients recovered, or where the Author had no Opportunity of examining the Bodies after Death.

§ 1. *Of the different Kinds of Cough.*

Cough without Expectoration; or with Expectoration of Mucus only.

THIS cough is commonly moſt ſevere at firſt going to bed, and is troubleſome by fits during the night; in ſome caſes, however, (though rarely) it is worſe in the day-time. It is accompanied with difficulty of breathing, ſometimes with hoarſeneſs, and often with pains in the cheſt; but theſe are ſeldom obſerved till the cough has been of ſome ſtanding. The fits of coughing frequently terminate with an expectoration of frothy mucus, which affords conſiderable relief. I have, however, known inſtances where that relief has taken place, ſeveral hours before the ſpitting began. But the moſt remarkable ſymptom attending this cough, and which indeed characteriſes it, is, the peculiar kind of fever. After one or two ſhivering fits, or after ſlight fits of coldneſs and of heat alternately, which come on in the morning, or a little after mid-day, (ſometimes

on

on alternate days only) the heat begins, and continues all the afternoon, and during the night, and then commonly terminates in profufe fweating. Sometimes there is no coldnefs nor fhivering, but a continued heat, which increafes after mid-day.

The pulfe is always quick, generally about a hundred in a minute, with almoft conftant head-ach, inceffant thirft, lofs of appetite, frequent retching, and fometimes faintnefs. This cough frequently is occafioned by expofure to cold or moifture. Delicate young women, efpecially when incautious, in thofe particulars, about the menftrual period, are very liable to it. It fometimes terminates favourably, but oftener in *phthifis pulmonalis*; and may therefore be reckoned the firft ftage of this difeafe.

Cough, with Expectoration of thick Matter.

This cough attacks alfo by violent fits, commonly in the night, fometimes in the morning. The expectoration (which generally begins fome weeks after the cough) is yellowifh, or greenifh, and is fometimes flightly ftreaked with blood. It is thick, vifcid, and mixed with a little frothy mucus; at times fœtid, and of a difagreeable putrid tafte. Its quantity is often not lefs than two or three pints in twenty-four hours, but it diminifhes towards the end of the difeafe.

The pains accompanying this cough are of two kinds; viz. acute pains in the fides, which frequently precede the firft attack of the cough, and are often fo violent as to ftop it; or forenefs, in the edges of the hypochondria, in the upper part of the recti abdominis mufcles, or in the loins; which follows the fits of coughing.

In

In ſome caſes there is no pain at any period; frequently there is fever, though it is ſeldom preceded by coldneſs and ſhivering, nor is it, in general, ſo regular as that which accompanies the cough firſt deſcribed. Sometimes, in the laſt ſtage, there is no fever, the pulſe being only ſixty in a minute. At this period alſo, purging, dropſical ſwelling, or profuſe ſweats take place; though ſometimes none of theſe ſymptoms occur during the whole courſe of the diſeaſe. As the cough abates, the fever, purging, and ſwelling abate alſo.

This cough is commonly produced by the ſame cauſes as the preceding, but ſometimes the cauſe is unknown. It frequently proves fatal in a few months; but ſometimes the patient is, for a number of years, ſubject to fits of it, which continue for ſeveral months at a time, eſpecially during the winter, and, in women, during pregnancy.

Cough, with Blood ſpit up in ſmall Quantities.

This ſpitting of blood commonly happens only in the more ſevere fits of coughing; it is preceded by violent pains of the cheſt, and accompanied with great difficulty of breathing, conſiderable fever, and ſometimes ſhiverings. The pains of the cheſt are, at times, increaſed by preſſure; when theſe, and the ſpitting of blood, come on without any evident cauſe, they are often removed, in a week or two; but when they attack after expoſure to dampneſs or cold, they generally terminate in ſpitting of matter, and a fatal phthiſis.

When theſe ſymptoms are occaſioned by an external injury, the ſpitting of blood ſeldom continues above a week, and all the complaints ceaſe in about a month; unleſs when it terminates in dropſy, which

which is ſometimes the caſe. As the ſpitting diminiſhes, it is more or leſs mixed with a yellowiſh matter, and at laſt becomes entirely purulent.

In ſome patients this complaint becomes habitual; continuing for many years, and attacking chiefly in the winter, or after any violent exertion.

Cough, with Blood flowing from the Mouth, by Fits.

Frothy blood is brought up by fits of coughing, which are, in ſome caſes, extremely ſlight; in others, are violent, immediately before the blood begins to flow; the quantity brought up at once is about half a pint, or a pint: it is generally pure, but ſometimes mixed with matter. The blood, in ſome caſes, flows only twice or three times during the paroxyſm; in others, much oftener. The approach of each fit is commonly known by the patient's expectorating more eaſily than uſual; and when coming on, the blood is felt riſing warm in the breaſt.

Theſe paroxyſms of hæmoptoe are ſometimes preceded by a cough of ſeveral months continuance, accompanied by an expectoration of matter, or of blood in ſmall quantity, or of a mixture of both; in other caſes they ſupervene a hoarſeneſs brought on by expoſure to cold.

This kind of hæmoptoe is accompanied by ſlight pains of the cheſt, (chiefly about the ſcrobiculus cordis) with faintneſs, heavineſs and drowſineſs, which ſymptoms are greatly increaſed before each paroxyſm, and are attended with conſiderable fever; the pulſe being

 ſometimes

ſometimes one hundred and thirty in a minute. Fits of coldneſs, and of ſweating, with ſickneſs, retching, and purging, are alſo not unfrequent ſymptoms of this complaint; which, for the moſt part, terminates fatally, though ſometimes in recovery.

A remarkable Inſtance of Recovery from a violent ſpitting of Blood.

A man, aged twenty-eight, had, for about a week, complained of pain and ſwelling at the pit of the ſtomach, and under both hypochondria; the pain was greatly encreaſed by the ſlighteſt preſſure, eſpecially on the right ſide, by lying on the left, or by a full inſpiration: it was accompanied with a trifling cough, but with a very high fever, the pulſe being commonly about one hundred and thirty in a minute. In this ſtate, he was ſeized with moſt violent fits of coughing, during which he ſweated profuſely, particularly on the upper parts of his body, and expectorated a conſiderable quantity of a thick, browniſh, red, ſmooth, or frothy matter. The cough and ſpitting having been almoſt inceſſant for thirty hours (ſome hours in the night only excepted), the pain of the right ſide, and the difficulty of breathing, decreaſed; the ſwelling diſappeared, the fever abated, and the pulſe fell to one hundred and eight; but, in about twelve hours, all theſe ſymptoms, except the ſwelling, returned with violence, the fits of coughing laſting, with hardly any intermiſſion, for three, four, ſometimes ten, and once for twenty hours at a time. The matter which he expectorated was often extremely fœtid, became gradually more bloody, and, at laſt, in the fits of coughing, he brought up pure blood in conſiderable quantity, and which ſometimes flowed from his mouth, uninterrupted by the cough; the pain and fever were always relieved after the fits of coughing and hæmorrhage, and they increaſed after theſe ſtopt or abated. About the four-

teenth

teenth day from the firſt violent attack of the cough, he began to ſpit, in ſmall quantity, a white matter ſtreaked with blood. The cough and fever now decreaſed very ſenſibly, the pulſe fell to one hundred, the pain went entirely off, the breathing became eaſy, and, in a fortnight, he had gained ſo much ſtrength as to be able to quit the hoſpital. In a fortnight afterwards, his complaint again returned with as much violence as before, and had nearly the ſame duration. Since this time four years have elapſed, during which he has never had the ſmalleſt complaint in the cheſt, and now enjoys perfect health and ſtrength.

§ 2. *Of difficult Breathing, or Aſthma.*

In this complaint the patients commonly breathe, with a wheezing or crackling noiſe, thirty or forty times in a minute, and ſtill oftener after eating, or after the moſt moderate exerciſe. They feel a general uneaſineſs in the upper part of the body, which commonly obliges them to ſit up; and likewiſe a tightneſs or pain acroſs the ſcrobiculus cordis, which prevents them, whether ſitting or lying, from ſtraightening the ſpine, and obliges them to keep the body much bent forwards; and ſometimes makes them lie with their knees drawn up. They complain of a ſenſe of weight either in one or both ſides of the cheſt, or at the pit of the ſtomach; when this laſt is the caſe, they ſometimes lie on their face, and when they turn on their back, have the ſenſation of ſomething falling from before; or if they turn to either ſide, of ſomething falling from the oppoſite ſide. They often awake in a fright. Their pulſe is about one hundred in a minute. This diſeaſe is not unfrequently attended by a cough with ſpitting, or by dropſical ſwellings; and ſometimes by rheumatiſm. It continues for many years, increaſing by fits; and I have not known it, when unaccompanied with other diſeaſes, prove fatal.

Is it not probable, that in theſe caſes there is a fluid in the cavity of the cheſt? or a ſuperabundant quantity of fluid in the pericardium? or that this membrane (in conſequence of inflammation) adheres to the forepart of the cheſt?

An inſtance of difficult Breathing relieved, upon ſoft Tumors appearing externally.

A woman, aged ſixty, formerly very healthy, after having been for ſeveral nights expoſed to cold, was ſeized with great pain and difficulty in breathing, and with a ſevere dry cough, from which ſhe was ſeldom free above an hour in the day. In about a month from the beginning of the complaints, a tumor appeared on the left ſide near the lower part of the ſcapula; and, a month afterwards, two ſmaller tumors were obſerved a little above the mamma of the ſame ſide: as theſe tumors increaſed in ſize, her complaints abated, and, in nine months, when the tumor on the back, now almoſt hemiſpherical, was larger than a new-born child's head, and each of thoſe on the breaſt nearly the ſize of an apple. She perceived no difficulty in breathing, unleſs after exerciſe, and her cough was ſeldom ſevere: ſhe had ſcarcely any pain in the tumors, which felt ſoft, as if they contained a fluid, and the ſkin which covered them was of the natural colour. When ſhe coughed, the tumors ſwelled, became hard, and, as ſhe imagined, were in danger of burſting.

In the preceding caſe, was there not an evident communication between the tumors and the cavity of the cheſt? Is it not probable, that

that matter, formed in the cavity, had made its way through the parietes? Could these tumors have been opened with safety, or advantage?

A Case of difficult Breathing immediately relieved, by the spontaneous Discharge of Matter from the Side.

A woman, aged twenty, received a violent blow with a man's fist on the lower part of the right scapula; she fell down instantly insensible, and remained so an hour: when she recovered her senses, she could hardly breathe, and the part where she had been struck was swelled and discoloured. Three days after the accident, she began to spit blood by coughing (sometimes in clots), and she continued to do so for two months, during all which time she could not endure any posture but laying on her face, resting on her elbows and knees. In about ten months, the pains in her chest, and difficulty of breathing, having nearly left her, her only remaining complaint being fits,* which came on soon after the accident, and to which she had been subject ever since, she was seized with chilliness, shiverings, cold sweats, sometimes partial, sometimes general, head-ach and giddiness, her pulse was about eighty-four, her skin itched violently, and many small itchy pimples, and painful blisters, appeared on it. After twelve months, the pain of the right side again increased; in fourteen months it affected greatly her breathing, and she could not bear even the gentlest exercise, nor lie on the right side; in fifteen months, she was obliged to sit up constantly, supported in bed, and frequently said, that something was collecting in her right side, although there was no swelling or discolouration of the part to be observed. Towards the end of the fifteenth month, a slight redness appearing

* Probably of the hysterical kind.

appearing in one part of the ſide, a poultice was applied, and, in a few days, matter burſt forth, to the quantity, as ſhe informed me, of two quarts; her difficulty of breathing was inſtantly relieved. Several months after this, ſhe had ſevere pains, or ſtitches, in her ſide near to the wound, which was between the ſixth and ſeventh rib; but theſe were removed by the application of a bliſter, and the appearance of many large boils. She has, ever ſince this time, though now four years after the accident, enjoyed perfect health.

§ 3. *Of Pains in the Side.*

Theſe pains are ſometimes ſo acute, and ſo much increaſed by inſpiration, that the patient dares hardly attempt to breathe. He cannot bear the ſlighteſt preſſure on the part affected, nor, while the pain continues violent, lie upon that ſide; but commonly lies on his back, with his head very high. His pulſe is ſmall, and about one hundred in a minute; with thirſt, and ſometimes head-ach. After the abatement of the pain, there is often a ſlight cough, without expectoration; and a degree of breathleſsneſs, after exerciſe. The patient ſometimes recovers perfectly in a few days, but ſometimes the complaint laſts from one to three weeks. Theſe pains, in ſome caſes, attack by fits, and then they are of longer duration; or they accompany hyſterical ſymptoms, but are then ſeldom fixed. The cauſe of them is frequently unknown; they ſometimes come on from expoſure to cold, and ſometimes are occaſioned by external violence.

C H A P. III.

Obſervations on the Effect of Remedies employed in the Cure of Diſeaſes of the Cheſt.

IN diſeaſes of the cheſt I have hardly ever obſerved any certain good effect from internal medicines. Vinegar of ſquills, has, on ſome occaſions, ſeemed to give relief to patients affected with cough and difficult breathing; and oily medicines, or ſpermaceti, appeared almoſt certainly to allay, for a ſhort time, violent coughing. But the remedies which have ſtill greater and more laſting effects, are bleeding, bliſters, and other local diſcharges; alſo fomentations. Bleeding is the appropriated remedy for a cough, and, except in the laſt ſtage of conſumption, ſeldom fails to afford very conſiderable relief, which ſometimes is felt immediately after the operation, at other times not till the next day, or even the third day; and upon ſome occaſions, not till after repeated bleeding. This remedy is alſo of ſervice in caſes of difficult breathing, and in pains of the ſide; although, for the latter complaint, the appropriated remedy is a bliſter, which almoſt conſtantly gives relief either immediately, or the day following.

Bliſters are likewiſe of conſiderable efficacy in caſes of difficult breathing, or hoarſeneſs, and ſometimes of cough. Setons or iſſues are uſeful in pains of the cheſt; and fomentations are of ſervice in pains occaſioned by external injury. From the early application of theſe remedies, pains of the ſide frequently, and dry coughs, ſometimes, terminate favourably; but if they are delayed for a week or a fortnight,

fortnight, the diſeaſe does not yield to them, but ſeems to keep on in its natural courſe.

In caſes of cough with expectoration, and of difficult breathing, or aſthma, theſe remedies ſeem to afford only a very tranſitory relief, and to contribute but little towards retarding the progreſs of the diſeaſe. Thoſe diſorders, therefore, which are the moſt common, and the moſt fatal of any, are unfortunately leaſt under the power of phyſic. I have known good air of ſervice in ſuch caſes, after bleeding had failed to afford even a temporary relief.

PART III.

PART III.

*Diseases of the Fluids.**

CHAP. I.

Diseases of the Fluids, illustrated by Dissection.

§ 1. *Extravasation of the Serum, or thinner Part of the Blood.*

A WOMAN, aged twenty-three, who had never menstruated, and, for many years, had been in a bad state of health, but without any particular complaint, became anasarcous about six weeks before her death. She had no cough, nor was her breathing laborious, although she frequently sat up in bed, and speaking was troublesome to her: she only complained of the swelling of her body, and of weakness of her eyes. She died quietly, and rather unexpectedly, in the night.

A transparent yellow fluid was found in most parts of the cellular membrane, a similar fluid in the abdomen, and a more than usual quantity of fluid in the pericardium, which adhered by a broad surface to the ribs and pleura of the left side: the lungs of this side, on first opening the thorax, were not visible, but after separating the pericardium from the ribs, and drawing it forwards, they were seen

H firmly

* I chose this title, not from any idea that the diseases described under it were diseases of the fluids only, and that the solids were not likewise affected; but because the changes which took place in the fluids were evident to the senses, whilst those of the solids were not.

firmly adhering to the posterior part. They were intirely red, not more than three inches broad, thin in proportion, not divided into lobes, not vesicular, every where hard, and, at each extremity, nearly of a tendinous consistence, and adhering so firmly to the ribs, that they could not be separated from them by pulling: the lungs in the right side adhered likewise to the parietes, in other respects were sound, as were the abdominal viscera.

As anasarcous swellings so frequently accompany diseases of the lungs, is it not probable, that they have some dependance on the state of this organ?

§ 2. *Extravasation of the red Part of the Blood.*

A woman, aged fifty, who, a fortnight before, had been seized with a fever, of which she could give no distinct account, complained (the fever still continuing) of pains all over her, and red spots appeared on her arms, breast, and legs: she was costive, her tongue parched, and covered with a black crust, or slough; her pulse, small, but not very quick; she was at times delirious, and often quite sensible. The colour of the spots becoming gradually darker, and her pulse sinking, she died about the nineteenth day of her illness.

Forty hours after her death, we examined the body, when, besides red and purple spots, of about a quarter of an inch diameter, which were very general on the surface, particularly of the extremities; there were also some blue blotches. The purple spots, viewed with a microscope, appeared of an uniform colour, whilst the red spots, which were broader, seemed a congeries of minute specks and ramifications. The cuticle, separated from the skin by boiling

water;

water; was found not in the leaſt affected by the ſpots, which appeared more diſtinct after its removal: they were confined entirely to the ſkin, though only viſible on its external ſurface; and they diſappeared altogether, when, after removing the cuticle, the ſkin was macerated in water. Sometimes, immediately under the ſpots, there were ſmall effuſions of blood in the cellular membrane; and, under the blotches, both this membrane and the fat were entirely red; but the muſcles never were affected. After having very carefully and ſucceſsfully injected one of the arms, no extravaſation of the injecting fluid could be perceived, either on the ſkin, or in the cellular membrane; nor could we ſee any extravaſation or enlargement of veſſels on a piece of the ſkin, which had ſeveral ſpots, when injected and dried. The viſcera of the thorax and abdomen were ſound: in the cavities of the heart and large blood-veſſels, were found ſmall but not firm polypi. A bit of the black cruſt, or ſlough, taken from the tongue and macerated for ſome days in water, tinged it red: what remained was a white mucus, readily diffuſible in water, and ſomewhat reſembling moiſtened bread.

As there was no extravaſation to be diſcovered after a very minute injection, is it not probable that the extravaſation of the colouring part of the blood, in the preceding caſe, was more owing to the ſtate of the blood than to that of the blood-veſſels?

§ 3. *Extravaſation of coloured Serum, &c.*

A man, aged twenty-four, was ſuddenly ſeized with ſhivering, which, after returning two or three times, was followed by the ſymptoms of fever; he complained chiefly of heat, ſlept much in the day; in the night was often delirious: his pulſe was quick and ſmall,

 his

his tongue dry, but not foul; he had two or three ſtools a day, and his urine let fall a copious cream-coloured ſediment, reſoluble by heat. On the ſeventeenth day, he felt a pain in the lower part of his thigh, which, on the twentieth, was greatly increaſed, and extended down the outſide of the leg; the parts affected were red, and ſomewhat ſwelled: his tongue was parched, and his pulſe fuller than before. On the twenty-fourth, the limb had become in part livid, the leg and foot were greatly ſwelled and painful; his countenance was pale, his tongue black, his pulſe fluttering: and, on the twenty-fifth, the day of his death, the cuticle was raiſed in bliſters, the leg having exactly the ſame appearance as it had two days after, when the body was examined.

The lower part of the right thigh was, on the outſide, red, or livid, and covered with ſmall bliſters, containing a red liquor; the lower part of the leg and foot, of the ſame ſide, were very much ſwelled, and, on the outſide, which was of the ſame colour with the thigh, there was a great bliſter, from which, before death, two ounces of a tranſparent red fluid, without ſmell, were taken. On the right arm, was a ſlight livid ſpot. In the veins of the pia mater, we found blood and air alternately interpoſed. The fluid in the lateral ventricles, was reddiſh, and coagulated ſlightly by heat. The liquor taken from the leg being immediately mixed, in different proportions, with water, gave it a red tint, the mixture remaining tranſparent; but, next morning, there was either a white cloud formed, a fur adhering, or a ſediment depoſited: the ſame liquor unmixed with water, ſuffered no change 'till the fourth day, when it alſo let fall a ſimilar ſediment; but it ſtill, and for ſeveral days afterwards, remained tranſparent: it was coagulated by heat almoſt as firmly as the ſerum, a ſmall quantity of an aqueous fluid remaining.

Did

Did not the firm coagulation, by heat, of the extravaſated fluid, ſhew that it was chiefly ſerum? Did not the admixture of the red part of the blood with this fluid, and with the lymph of the lateral ventricles, ſhew a tendency to putrefaction? although, as there was no diſagreeable ſmell, this could hardly be ſaid to have taken place. Was not the air interpoſed between the portions of blood in the veins of the pia mater, a further evidence of this tendency? for I have, on other occaſions, remarked, that large veſſels which paſſed near an internal putrid ulcer, contained, in like manner, portions of blood, and of air, alternately.

§ 4. *Putrefaction of the Fluids.*

A woman, aged twenty, was ſeized with ſhiverings, followed by fever; ſhe became dull, heavy, ſtupid, and ſometimes delirious: ſhe had a violent purging, her tongue and eyes were parched, her pulſe quick and ſmall, and there were petechiæ on the right arm; ſhe was quite neglected 'till the tenth day of her illneſs; ſhe died on the eleventh, and, immediately after death, a change of colour took place in the body.

This diſcolouration was principally on the right ſide, from the breaſt to the middle of the thigh, and from the linea alba to the ſpine; the upper and lower parts, and belly, were green, the back livid, and the pudenda quite black: at inciſions made on any of thoſe parts, a conſiderable quantity of a muddy liquor, nearly of the ſame colour with the part, run out: it was ſo intolerably fœtid, that a man had almoſt fainted from ſmelling to it; the parts from which it flowed were tender, and eaſily pulled aſunder. In the pudenda, the blackneſs penetrated to the cellular ſubſtance

and

and fat, but it did not extend inwardly beyond the nymphæ, and backwards not quite to the anus; the green colour of the abdomen penetrated through the integuments, the fat, and the oblique mufcles; but the recti, tranfverfe mufcles, and peritoneum, were free from it. The livid colour of the back penetrated almoft to the bones, near which the mufcular fibres appeared found; there were a few red fpecks on the arm and breaft of the right fide; nothing preternatural appeared in the cavity of the abdomen, except one black fpot on the fundus uteri.

After having vifited this woman, I became for a minute, blind, ftupid, and confufed, but I fuffered no inconvenience either during the diffection, or afterwards. One drachm of the putrid liquor received at the incifions, was, an hour afterwards, along with three drachms of water, injected into the crural vein of a healthy bitch, who was giving fuck: in a minute fhe vomited; in an hour all her limbs trembled, and in an hour and a half fhe feemed in the greateft uneafinefs, whilft her puppy, who had given over fucking, was making a noife: fhe frequently vomited during the day, and in the night; next day, when called to, fhe moved flowly and feebly, and could hardly keep her eyes open; her hair ftood on end, and fhe refufed taking food 'till the evening, from which time fhe gradually recovered.

Twelve hours after death, I examined the body of a young man who died of a fever refembling the preceding. The fkin of the left breaft was brown, and the pectoral mufcle had loft its colour and was rotten; the liver, likewife, was in fo tender a ftate, that a very fmall force was requifite to pufh the finger into any part of it. The other abdominal vifcera appeared found.

Soon

Soon after examining this body, I felt an acute pain at the end of the finger which I had puſhed into the liver; it inflamed: a ſmall piece of it near the nail became black and mortified, and, after a few days, was thrown off by ſuppuration.

Are dullneſs, ſtupor, and lowneſs of pulſe, the criteria of this fever, or of a tendency to it? Is infection more readily communicated from the living than from the dead body? May an external part of a healthy body be affected without injuring the reſt of the ſyſtem? If the putrid matter has been mixed with the blood, will a putrid fever always follow? Is there a connection between this and the petechial fever? From the external parts being principally affected, is it not probable that the air has ſome influence in promoting the putrefaction? Would not the external application of antiſeptics have an effect in retarding this proceſs?

§ 5. *Extravaſation and Putrefaction, united in the ſame Subject.*

A woman, aged fifty, who, though addicted to the uſe of ſpirituous liquors, was healthy till four years before her death, when ſhe was ſeized with pains of the cheſt, cough, and difficulty of breathing, which going off in a few months, ſhe continued well for three years: her complaints then returned, and, for two months, were accompanied with a very copious diſcharge of thin ſaliva; ſoon after the ſtopping of which, ſhe became generally œdematous. During a few months in the ſummer, the ſymptoms were moderate, but, four months before her death, they were more violent than ever; her breath being very ſhort, obliged her to ſit up almoſt conſtantly; for, when lying, ſhe was in danger of being ſuffocated; ſhe could utter but a few words without ſtopping; her cough was very troubleſome, and, in the fits of it, her face, which was at all times bloated,

bloated, became purple; ſhe often performed all the motions of coughing, without uttering any ſound; ſhe was thirſty, her tongue dry and whitiſh; her pulſe about one hundred; her urine depoſited a copious ſediment, like powdered bark, the liquor above reſembling diluted claret. The ſwelling was univerſal, but the lower extremities were ſo much diſtended, that they could hardly be moved. The belly being alſo greatly diſtended, ſhe was tapped, and ſeveral gallons of a greeniſh liquor, by this means, were evacuated. After the operation, ſhe was, for ſeveral days, much relieved; but, on the ſixth, having had a violent fit of ſhivering, ſhe died ſuddenly.

In the month of February, thirty-four hours after her death, I examined the body. It was generally ſwelled, with red ſpecks and purple ſpots on ſeveral parts of the ſurface; the latter, about one third of an inch in diameter, penetrating quite through the ſkin, were owing to extravaſated blood. The face was of a deep purple; on the breaſt were long, ramifying, red vibices; a portion of the cuticle, on the inſide of the thigh, was raiſed in a bliſter, containing a thin, reddiſh fluid, and, for a conſiderable way round the bliſter, was eaſily ſeparable from the true ſkin: in the abdomen, were ſeveral quarts of a reddiſh muddy fluid, in ſome degree coagulable; the contents of this cavity were ſound; the liquor pericardii was reddiſh; the lungs every where adhered, by a cellular ſubſtance, to the neighbouring parts, and, though there was no particular hardneſs, or tubercle, to be ſeen, they were not quite ſo ſoft as in a ſound ſtate; air blown in at the windpipe eſcaped, and diſtended the cellular ſubſtance on their ſurface; the ſame thing happened, when, in a portion of the lungs cut off, air was blown in at a branch of the windpipe, or pulmonary vein: when the pulmonary artery was blown into, its ſmaller branches were alone diſtended; a black ſubſtance accompanied he veſſels in their courſe, and appeared, upon a tranſverſe ſection, like a black circle ſurrounding them.

Was the want of ſound in coughing owing to the air, in the time of that action, eſcaping from the veſicles into the cellular ſubſtance? Is this, therefore, a ſign of ruptured veſicles? Did the copious diſcharge of ſaliva, in the beginning, ſhew a ſeparation of the thinner fluids; which, upon the ſtopping of that diſcharge, were depoſited in the cellular membrane, and in the abdominal cavity? Did the purple colour of the countenance, the ſpots and vibices, ſhew that the red part of the blood was broken down, and entered veſſels which, in a ſound ſtate, it could not do? Did the colour of the urine, of the liquor in the bliſter, abdomen, and pericardium, ſhew a ſolution of the red part of the blood in the thinner fluids? Do the fluids in cavities ſometimes acquire a red tint, either a few days before, or a few days after death? Was the diſeaſed ſtate of the fluids, in the preceding caſe, to be imputed to the affection of the lungs, to the uſe of ſpirituous liquors, or to both?

In diſſecting the body laſt mentioned, having broken the ribs that the lungs might be more fully ſeen, the pointed ſplinters of them punctured the cuticle on ſeveral of my fingers, eſpecially of the left hand. On returning home, about an hour after, I was ſeized with ſhiverings, wearineſs, and pains all over my body: towards evening I had a violent head-ach, and ſome degree of fever; but next morning awoke perfectly well. The wounds did not heal up, but became red, ſwelled, and, though not very painful, for ſeveral months gradually increaſed. Having the ſame day, immediately after examining this body, examined the body of a man who had had a veneral complaint, it was ſuppoſed that from the latter, the injury might perhaps have been received: I, therefore, made trial of a variety of mercurial applications, after each of which the tumors became larger, redder, and more painful; they were then burnt down by lunar caustic, but always grew up again, and, at last, had the appearance of warts, with their bases swelled and red. In less than a twelvemonth after the accident, there appeared on the back of the left hand, where there had been no wound, a small, moveable, round tumor, resembling a lymphatic gland, which gradually increased; the skin became livid, and there were sometimes slight shooting pains in the part. Soon after the appearance of this tumor, the glands in the left armpit swelled, and became, in some degree, painful; the glands of the same side, under the lower jaw, also swelled, and several small sores broke out upon the tongue and inside of the mouth; soon afterwards the glands also of the right armpit swelled. Being at a loss how to stop the progress of so uncommon an affection, I began to take mercurial medicines, in considerable quantity, which I had several times employed before, though sparingly: at last, I underwent a salivation by unction; but never observed, during this course, nor for a considerable time after it, the smallest favourable change upon the tumors on the hands, the swelled glands in the armpits, or the sores in the mouth; I, therefore, submitted to have the tumors removed by the knife, several of which growing up again, were repeatedly cut off. The sores, while healing up, never had a good appearance, but were sloughy, and sometimes very painful. In about a fortnight, after the last remaining tumor had been cut off for the second time, and about two years after the wounds had been inflicted, the swelling of the glands subsided, the sores in the mouth healed up, and have not (though it is now almost three years) in any degree returned.

CHAP. II.

C H A P. II.

A Deſcription of the Symptoms of the Diſeaſes of the Fluids, taken from thoſe Caſes where the Patients recovered, or where the Author had no Opportunity of examining the Body after Death.

§ 1. *Swelling of the Belly, with Fluctuation.*

THIS ſwelling ſometimes occurs alone, but moſt commonly it is attended with external, or œdematous ſwellings. Women are chiefly ſubject to it from obſtruction in their menſes, or during pregnancy, in which caſe it continues after delivery. It happens to perſons of either ſex after a cold, cough, or fever, and ſometimes without any diſeaſe preceding it. It is accompanied with difficult breathing, cough, thirſt, a diminution in the quantity of urine, and, at times, with fever. The patients, in general, have a faded, or ſallow complexion, though ſometimes they retain a ruddy and healthy appearance. I have known this complaint continue five years, without cauſing any conſiderable uneaſineſs; but when the ſwelling has reached its utmoſt extent, the condition of the patient is truly miſerable: obliged to lean forward, the belly ſupported by pillows, tormented with violent pain in the bowels, and with bile forced up into the mouth almoſt every minute. This diſeaſe is frequently relieved, or carried off, at leaſt for a time, by ſpontaneous purging; ſometimes by ſpontaneous ſweating. When it returns, as it ſometimes does, ſeveral times in a year, it terminates fatally.

Might not ſudorifics be tried in thoſe caſes where purgatives have failed?

§ 2. *General external Swelling, retaining the Impreſſion of the Finger.*

This ſwelling is more remarkable in the lower, than in the upper parts of the body, and is frequently more conſiderable in the right than in the left ſide; the parts affected, are, ſometimes, though rarely, tender and painful to the touch. This complaint is ſometimes unpreceded by any other, or it follows after ſickneſs and indigeſtion, ſudden ſuppreſſion of the menſes; and frequently after a cough, or ſome other affection of the cheſt. It is commonly accompanied by difficult breathing, thirſt, and paucity of high-coloured urine, becoming turbid when cold. Perſons of very different ages, from twelve to ſixty-two, are ſubject to this diſeaſe: ſome have the countenance bloated, others have a ſpontaneous and very conſiderable bleeding at the noſe; in either caſe, and not otherwiſe, the diſeaſe terminates fatally.

§ 3. *General external Swelling, wlth Swelling of the Belly.*

The union, or combination, of theſe complaints, is to be met with moſt commonly in perſons naturally of a weak or unhealthy conſtitution, the external ſwelling almoſt conſtantly precedes the ſwelling of the belly; they follow from the ſame cauſes, as when ſingle or alone. I have known them happen, and to a high degree, in a fortnight, after a bruiſe on the cheſt. They are accompanied by the ſame ſymptoms as § 1. and § 2. often with faintneſs and lowneſs; ſometimes with vomiting of blood, or purple ſpots on the

the ſkin, both fatal ſymptoms. They ſometimes terminate favourably by a ſpontaneous purging, increaſed diſcharge of urine, or a flow of thin fluid from the ſalivary glands.

Is not either ſpecies of dropſy commonly a ſecondary diſeaſe?

Is it not evident, from the bleeding at the noſe, vomiting of blood, and purple ſpots, that the red part of the blood is broken down, or the blood-veſſels weakened?

Should not the view of the phyſician be directed rather to the amendment of the fluids and ſolids, than to the evacuation of the former?

Is it likely that this amendment may be attained by animal food and Peruvian bark?

When the evacuation of the fluid is neceſſary, and when purgatives, diuretics, and ſudorifics have failed, might we not imitate nature, and excite a ſalivation?

§ 4. *A fluctuating Swelling on the Loins.*

A man, aged thirty-two, having been thrown down on his face, the narrow wheel of a loaded cart went directly acroſs his loins from left to right; he was carried home, neither wounded nor in pain; and, though benumbed in the loins, he walked in the evening. For ſome days he complained of pain in his bowels, and had no ſtool. The day after the accident, we obſerved a fluid collected under the integument of the loins. In a fortnight, the integuments were greatly ſwelled,

ſwelled, and an evident fluctuation was felt on ſtriking the tumor. In a month, the fluid having ſpontaneouſly decreaſed, did not fill the cavity which it had formed, and by change of poſture, or upon preſſure, it moved from place to place, the patient himſelf being ſenſible of its motion, as he had been before of its fluctuation. The cavity, at this time, extended from the os coccygis, ſeven inches upwards; and from the left great trochanter to within a few inches of the ſame tuberoſity on the right ſide. In two months, the fluid had almoſt wholly diſappeared, and the integuments having become firm, adhered to the parts underneath. In a few weeks more, he was free from complaint.

N. B. Fomentations had been employed.

§ 5. *Fever, with red, or purple Spots on the Skin.*

This diſeaſe commonly begins with a fit of ſhivering, which ſometimes returns, and is always followed by fever. The patients, in general, take to their beds in the beginning, although they ſometimes go about for three or four, or even nine days, uncertain of the nature of their complaint. The ſpots appear, at different times, from the fifth to the tenth day, and are either very ſmall, of a deep red or purple colour; or larger, about one quarter of an inch in diameter, and of a paler red: both kinds frequently appear in the ſame perſon; the former chiefly on the extremities, the latter all over the body. The breathing is laborious, and ſometimes accompanied with a ſnorting noiſe. The eyes are red; the tongue and lips parched, or chopped, and covered with a black, tough, ſemi-tranſparent cruſt; which, by maceration in water, becomes at firſt gelatinous, and afterwards of the conſiſtence of ſyrup. Sometimes blood is effuſed on the tongue,

and

and hardens on its ſurface. The pulſe ſeldom exceeds one hundred in a minute, ſometimes ſlower; it is low, and ſtrikes the finger ſo gradually, that it ſeems rather to preſs upon than ſtrike it: it frequently intermits. The patients complain of pain, or of noiſe in the head, and of general uneaſineſs; they are dull, and ſo drowſy, that they can hardly keep their eyes open; they are ſometimes ſenſible, but more commonly delirious, eſpecially in the night, when, unleſs prevented, they frequently get out of bed, but are not outrageous, and are eaſily prevailed on to return again to bed. The liquor diſcharged by bliſters, though of a dark brown colour, is free from ſmell. The ſtools are ſometimes of the ſame colour, and the urine depoſits a ſpongy, ſometimes reddiſh ſediment, in ſmall quantity. there is ſeldom any tendency to ſweating, but very often to purging, the ſick having commonly four or five ſtools in twenty-four hours; and it is difficult to determine, whether this evacuation be hurtful, or ſerviceable. An abatement, or relief of the ſymptoms, commonly happens on the fourteenth day, ſeldom ſooner; frequently not till about the twentieth. For ſeveral days after the abatement of the fever, the patients are often troubled with a dry cough, and ſometimes become deaf. Of ten patients whom I ſaw in this fever, two died: the firſt, who had a moſt violent purging, died on the ſixth; and the ſecond, who was coſtive, died on the twenty-ſecond day. This fever appeared, in ſome inſtances, to have been communicated by infection.

Did the ſpots, the blood effuſed and concreted on the tongue; the colour of the liquor of the bliſters; that of the ſtools, and the ſediment of the urine; ſhew that the red part of the blood was broken down?

C H A P. III.

CHAP. III.

Obſervations on the Effects of Remedies given in the Cure of Diſeaſes of the Fluids.

§ 1. *A general Account of thoſe Effects.*

IN the ſpotted fevers, inſtances of recovery were moſt frequent after the uſe of bark, of cordials, and of bliſters. The ſwelling of the belly, or of the external parts, in general, ſubſided, though commonly only for a time, from the uſe of the more powerful purgatives; viz. jalap, elaterium, and dried ſquills, given either ſeparately or combined. Alſo during the uſe of diuretics, as nitre, the ſal diureticus, infuſion of horſeradiſh, and tincture of cantharides. The ſwelling of the belly conſtantly returned after tapping, and ſometimes the patients died very ſoon after this operation. The puncturing the legs was likewiſe attended with danger, and, in one caſe, the limbs inflamed, became black, and the patient died in three weeks. As evacuations then are moſt commonly ineffectual, and even dangerous, a medicine is greatly wanted which would produce ſuch a change in the parts, as after abſorption to prevent the further extravaſation of the fluid. The two following ſingular inſtances of the happy effects produced by mercury and bark, may poſſibly ſuggeſt ſome uſeful hints on this important ſubject.

§ 2. *The Effect of Mercury in an obſtinate Swelling of the Limbs.*

A young woman having, on the day her menſes began to flow, taken imprudently, whilſt hot, a draught of cold water, the diſcharge

immediately

immediately ſtopped; her legs inflamed and ſwelled; and ſhe was ſeized with ſhiverings, followed by fever and pains all over her body; after two or three weeks, the fits of ſhivering, ſucceeded by fever, frequently returned again, and at thoſe times the inflammation of the legs increaſed. In about a year, the whole of the lower extremities had attained an enormous ſize; but the ſwelling was not now attended with inflammation; on the contrary, the limbs were cold and hard: it differed alſo from the œdema, as it did not retain the impreſſion of the finger; nor was ſenſibly increaſed towards evening. In every other reſpect, excepting a diminution of the catamenia, the patient ſeemed to be in perfect health. Notwithſtanding the uſe of various medicines, and the application of cauſtics and bliſters, which laſt, by the bye, occaſioned no diſcharge, her limbs remained in the ſtate above deſcribed for almoſt two years and a half; ſhe then began to rub on her legs the mild mercurial ointment, gradually encreaſing the quantity to half a drachm, afterwards to one drachm every night; ſhe lived low, and the limbs were kept in a horizontal poſture. In three weeks, the ſwelling having ſubſided, the legs were ſoft and flaccid; and, in three months, the ſkin was ſo looſe, that it ſeemed probable, that what had formerly diſtended it, was now moſtly abſorbed. Her mouth was but little affected; her bowels not in the leaſt: ſhe ſweated much, and made water in conſiderable quantity.

§ 3. *The Effect of Bark in a copious Diſcharge of putrid Saliva.*

A woman, aged thirty-eight, after ſeveral irregular fits of coldneſs, ſucceeded by heat and ſweating, was ſeized with a ſwelling of her cheeks, which externally were tenſe and ſhining, and internally were covered with a firm white cruſt, or ſlough, above half an inch thick, and which was totally inſenſible when cut through with a knife; the palate and gums were alſo covered with a ſimilar ſlough; the tongue was almoſt unmoveable, ſo that her ſpeech was ſeldom intelligible, and the teeth could not be brought in contact, owing to this cruſt projecting between them: there was, day and night, a continual diſcharge from the mouth of a viſcid, ropy fluid, frequently bloody, and ſo abominably fœtid, that it was diſagreeable to come within ſeveral yards of her; the quantity diſcharged was about four pints in twenty-four hours. The fluid ſpontaneouſly ſeparated into two parts: the one a thinner; the other a more viſcid and heavier; the former not at all, the other only partially coagulable by heat; the patient could ſwallow only the thinneſt liquids, and even theſe with difficulty: ſhe had no appetite, but had daily four or five looſe ſtools: her pulſe was very ſmall, between one hundred and ſix, and one hundred and twenty. To this miſerable condition ſhe had gradually arrived; when, on the twelfth day of her illneſs, ſhe began to uſe the bark in the following manner:

R. decoct. corticis peruv. unciam unam.
Tinct. ejuſdem, drachmas duas.
Quarta quaque hora ſumendas.

In four days the diſcharge was thinner, and leſs offenſive; ſhe had leſs difficulty in ſwallowing, had ſome appetite; her pulſe was ſlower, her purging had ſtopped, and the cruſt began to fall off from the angles of the mouth; continuing to recover, ſhe, in a few days more, began to have pain in her cheeks, which before were inſenſible; afterwards in her tongue; and, laſtly, in her gums and teeth; the agony of which was ſometimes ſo great as to prevent her ſleeping. When the cruſt had fallen off from the gums, they appeared puſhed out in the form of papillæ. On the twelfth day, after ſhe began to take the bark, her face, outwardly, had almoſt its natural appearance; the cruſt had fallen off entirely from the cheeks,

palate,

palate, and gums, the papillæ of which had alſo ſubſided, and there was but little of it now remaining at the root of the tongue: ſhe could bring her teeth cloſe together, could move the tongue a little, and ſpeak more diſtinctly. The diſcharge, now far leſs copious and leſs fœtid than formerly, did not flow conſtantly, but, being collected in the mouth, was ſpit out; ſhe had hardly any pain, could ſwallow ſolid food, was coſtive, and her pulſe between eighty and ninety. In a few days more the diſcharge ſtopped entirely, and, in a few weeks, ſhe had no complaint but ſtiffneſs in the parts, which prevented her from opening her mouth, or putting out her tongue freely; and ſome months afterwards, being perfectly well, ſhe went into the country.

PART IV.

Diseases of the Head; Nerves, and Muscles.

CHAP. I.

Diseases of the Head, &c. illustrated by Dissection.

§ 1. *Lymph lodged between the Dura and Pia Mater.*

A STOUT man, aged twenty-three, was suddenly seized with a fever, accompanied almost constantly with violent delirium. When brought to the hospital, he walked about nearly as a person in health, and answered some questions sensibly; but the answers he gave to others, and a certain wildness in his look, discovered the dangerous state he was in. As he could not otherways be kept in bed, he was bound with broad straps of leather; his face and eyes were red, his pulse neither quick nor full; he always said he was very well, that his giddiness had gone off, and complained of his confinement. Every night, and frequently during the day, he was noisy and ungovernable, tearing whatever came within his reach. On the night but one preceding his death, he broke the straps, got out of bed, and it was with difficulty that three men could again lay him down. He then sweated profusely, his pulse sunk, his face became pale, his voice changed to a doleful cry; and, in a few hours afterwards, on the morning of the ninth day, he expired.

A woman,

A woman, aged ſixty, was ſeized with a fever, accompanied alſo with violent delirium: when aſked how ſhe did, ſhe commonly anſwered ſhe was very well, excepting towards the end, when ſhe complained of her head; ſhe was noiſy in the evening, and during the night, and it was with difficulty that ſhe could then be kept in bed. Her pulſe was quick, and ſometimes intermitting; her eyes red, her lips black: at laſt, her face having been frequently diſtorted, and her arms ſometimes convulſed, ſhe died on the thirteenth day.

A ſtout man, aged thirty-two, after ſeveral fits of ſhivering, was ſeized with a fever, accompanied with violent delirium, though leſs conſtant than in the preceding caſes: he was ſometimes ſenſible during the day, and complained greatly of his head; he was outrageous on particular nights only; his eyes were red, and his pulſe quick. He died on the eighteenth day.

In the three preceding caſes, after the ſcull had been ſawed through, on cutting the dura mater* and inclining the head, a thin colourleſs liquor, not coagulable by heat, run out, to the quantity of about two ounces, in the firſt caſe; of about an ounce and a half in the ſecond; and of about half an ounce in the third. After the upper part of the ſcull had been ſeparated from the dura mater, on raiſing this membrane, a ſmall quantity of lymph was obſerved lying between it and the pia mater; and, after the brain had been removed, the medulla oblongata and ſpinal marrow appeared ſurrounded with lymph. In the firſt caſe, there were no other preternatural

* We were led to perform this part of the diſſection with caution, having formerly found lymph in the encephalon of a man, who before death had ſimilar ſymptoms, but whoſe head was not opened ſo carefully as to enable us to give an exact deſcription of it.

natural appearances in any part of the encephalon; in the ſecond and third caſe, there was a ſmall quantity of lymph effuſed between the convolutions of the brain, under the pia mater; and, in the ſecond caſe, there was rather more than the common quantity of lymph in the lateral ventricles.

Are outrageouſneſs, and inſenſibility to pain, characteriſtics of this ſpecies of fever in its higheſt degree? From the violence of thoſe ſymptoms being nearly in proportion to the quantity of lymph effuſed, is it not probable that they are occaſioned by the preſſure of that liquor upon the brain?

§ 2. *Suppuration of Part of the Dura Mater.**

A woman, aged thirty-four, ſtooping to avoid a beam of wood which a man carried on his ſhoulder in the ſtreet, was ſtruck by it in the upper and fore part of her head ſo violently, that ſhe fell backwards: ſhe, however, got up again immediately, and, after wiping off ſome blood which came from the wounded part, walked ſeveral miles: ſhe alſo next day walked ſeveral miles, but complained that the cold ſtruck like a knife through her head. On the fourth day, ſhe came to the hoſpital, when a tumor, which had riſen about the wound, was laid open, and a fracture ſearched for, but none was found. Though ſenſible, ſhe ſometimes ſtarted and looked oddly. In the evening a delirium came on, and ſhe was noiſy a great part of this, and for many ſucceeding nights. During the day alſo ſhe was frequently inſenſible and delirious, eſpecially on the tenth and the three following days; ſhe often complained of exquiſite pain in her forehead, which ſhe would not ſuffer to be touched; ſhe alſo complained of weight and oppreſſion,

* *Fig.* xi.

preſſion, and always cried out at the dreſſing or enlarging of the wound: her face was often diſtorted, and her limbs convulſed; ſometimes ſhe trembled, and frequently ſtarted when ſpoken to: the pulſe was ſeventy, and pretty regular. On the fourteenth day the delirium had greatly abated; ſhe became almoſt conſtantly ſenſible of her pains, and was apprehenſive of danger; on hearing the leaſt noiſe ſhe ſtarted, and ſaid it was like thunder in her ears: the pulſe was between ſixty and ſeventy. On the morning of the ſeventeenth, ſhe had a fit of ſhivering, which returned afterwards, four or five times, at irregular periods: it was followed by fever and ſweating: the pulſe, during the fever, varied from one hundred, to one hundred and thirty, according to the violence of the fit, and the ſhortneſs of the interval between it and the time of examining the wound. About the twenty-firſt the delirium went off entirely; ſhe was dull, drowſy, kept her eyes ſhut, and ſeldom ſpoke; ſhe afterwards became gradually weak, apt to faint when ſitting up to have her wound dreſſed: the pulſe ſunk; and, on the twenty-eighth day, after receiving the blow, ſhe expired.

The contiguous ſurfaces of the diſeaſed part, both of the dura mater and of the cranium, were each of them moiſtened with a little purulent matter; but the bones were ſmooth, and, in every reſpect, ſound. The internal ſurface of the dura mater, oppoſite to the diſeaſed part; the falx, and the two lower ſides of the longitudinal ſinus, appeared quite ſound, but the internal ſurface of the upper ſide of that ſinus, was of a light ſtraw colour, ſmooth, and, when held oppoſite to the light, did not ſeem thickened. All the other parts of the membranes of the brain, and of the cerebellum, were carefully examined, but we met with no other preternatural appearance.

Where

Where the reflexion of the dura mater forms the falx, is there any cellular ſubſtance in which pus may be formed and collected? Does matter collected there, make its way more readily through the external lamina of that membrane to the bone, than through its internal lamina to the brain? In the preceding caſe, was there no matter formed till about the ſeventeenth day, at which time the delirium went off, and the ſhiverings began, which were ſoon followed by drowſineſs? If fatal conſequences ariſe from a few drops of matter in the head, what have we to apprehend from a larger quantity of this, or, probably, of any other fluid, collected there.

§ 3. *The Veins of the Pia Mater apparently enlarged.*

A man, aged twenty-three, after having been drunk and riotous in the ſtreets for two nights, was ſeized with pains, particularly in his head, attended with fever: he ſweated in the beginning. On the fourth, and ſome following nights, he was delirious, though not unmanageable: during the day he was ſenſible, until the ſeventh, after which he appeared quite ſtupid, and his eyes, both day and night, were open and fixed; his body at firſt coſtive, was now looſe, and his ſtools and urine came away involuntarily; he ſweated profuſely, and died quietly on the eleventh day.

In ſawing the ſcull rather haſtily, the veins of the pia mater, which terminate in the longitudinal ſinus had been, as we afterwards found, cut through, and about two ounces of blood had run out; yet thoſe veins, when the cranium and dura mater were raiſed, appeared ſtill diſtended with blood, and greatly enlarged. The other parts of the encephalon were examined, but nothing preternatural was ſeen. The lungs and heart were in a ſound ſtate.

§ 4. No

§ 4. *No morbid Appearance in the Brain after an Apoplexy.*

A ſtout healthy man, aged thirty-one, who in the morning went out perfectly well, in the courſe of the day complained of giddineſs and head-ach. In the evening, when walking with ſome companions, he ſtopped ſuddenly, ſaying, that his head-ach and giddineſs were ſo great he could not go on. He immediately leant to a ſide, was violently convulſed, and, though his friends ran to his aſſiſtance, fell to the ground ſpeechleſs and inſenſible. Twenty-four hours after the attack, he lay on his back, breathed high, with a ſnorting noiſe; his face was turned to the right, his mouth and noſe drawn a little to the ſame ſide, and froth iſſued from his mouth; his eyes were ſhut, whilſt tears ran from them; his pulſe was quick and very high; his ſkin hot, and moiſtened with ſweat; and the muſcles of the arms quite relaxed. After thirty-ſix hours, there was little alteration in the ſymptoms, only that the eyes were now open, though fixed, and not affected by the near approach of objects; pulſe alſo was lower, and his ſkin was neither ſo hot, nor ſo moiſt as at firſt. After forty hours, the alæ naſi were during inſpiration drawn almoſt cloſe to the ſeptum; and, indeed, the trunk of the body ſeemed drawn up at the ſame time: the right ſide of the face was now frequently diſtorted; the right arm ſometimes convulſed: the heat of the body diminiſhed; the pulſe was no longer to be felt, and, in forty-five hours, he expired.

The membranes, ſubſtance, and ventricles of the brain, cerebellum, and medulla oblongata, were all carefully examined, but nothing preternatural was obſerved. Upon inverting the trunk of the body, about an ounce, or more, of a bloody fluid, ran out from the ſpinal canal. The ſpinal marrow, owing to ſome neglect, was not examined. The viſcera of the thorax, and of the abdomen, appeared perfectly ſound. The ſtomach contained a yellow fluid.

§ 5. *Deep red Blotches, and partial Suppuration in ſeveral Muſcles of the Body, in Conſequence of a Wound.*

A man, aged forty-five, fell from a height of about eight feet, and ſtruck his left leg againſt the point of a pick-ax. He walked a good deal after the accident, and the wound, for ſeveral days, was quite neglected. On the eighth day, he firſt felt pain in his throat, and, in the evening, ſwallowed with difficulty. On the eleventh day, the lower jaw was fixed ſo cloſe to the upper, as to prevent him taking ſolid food. On the thirteenth day, when he was brought to the hoſpital, he could endure no poſture but that of lying prone, reſting upon his knees and elbows, raiſed up by pillows. Whenever he attempted to change that poſture, and very often at other times, he was violently convulſed, and ſometimes thrown out of bed. In the more moderate fits of convulſion, the courſe of which could be obſerved, he firſt ſtarted ſuddenly from the prone poſture to his knees; the body was then drawn forward, the head backward, and the lips ſtrongly preſſed together; though he was, at all times, careful to inſert the points of his fingers between them, over which, in breathing, the air ruſhed with a hiſſing noiſe. During theſe fits, which continued ſome minutes, the recti abdominis, ſterno-maſtoidei, and other muſcles on the fore part of the neck; thoſe within the arch of the lower jaw, and the maſſeters, were hard to the touch. He told us, that the fit began with a catching below the navel, that he had, at all times, exquiſite pain in that part, and alſo in the fore part

part of the neck, and near the jaw, and that he ſhould be ſtifled, unleſs he kept his lips aſunder, by inſerting his fingers between them. The lower jaw was always immoveably fixed ſo cloſe to the upper, that the point of the finger could not be puſhed between the teeth, and any interval between them was more owing to the lower jaw being drawn back than to its being depreſſed: the ſweat ſtood in drops upon his face and his body; his pulſe was ſmall, and between eighty and ninety: though he had the ſtrongeſt deſire to drink, yet the great difficulty he found in ſwallowing, made him moſt reluctantly put the cup to his head, and frequently withdrew it; and, when he had ſucked in a little, he only ſwallowed a ſmall part of it; the reſt was ſquirted out immediately, and the lips and muſcles of diglutition were ſtrongly convulſed. In the evening of the fourteenth day, his face was pale, his body covered with cold ſweat; his pulſe intermitted; and when his convulſion fits came on, he could not raiſe his hand to keep his lips aſunder, which was done by an aſſiſtant: the lower jaw was conſiderably relaxed. In the morning of the fifteenth day, at three o'clock, he ſpoke ſenſibly: at four he was ſuddenly convulſed; and, though lying on his belly in bed, was thrown on his back upon the floor, and died inſtantly. Four hours afterwards, the body being ſtill quite warm, the lower jaw was immoveably fixed to the upper.

On the outſide of the tendo achillis of the left leg, there was a wound, which paſſing before that tendon, penetrated as far as the ſkin on the oppoſite ſide behind the internal maleolus. The cavity of the wound, which contained pus, and two bits of woollen ſtocking, was about the ſize of a pigeon's egg. The poſterior tibial nerve, in paſſing along it, was covered by a thin cellular ſubſtance, which, in the wound and ſeveral inches above it, was of a bright red; but the nerve cut acroſs appeared ſound. The muſcles which formed the

 ſides

ſides of the wound, were partially ſuppurated, and, for a conſiderable way up the leg, of a deep red. The cellular ſubſtance and tendinous aponeuroſis on the outſide of the leg were, as high as the knee, of a deep red; the latter unequally ſo, being in ſome places almoſt black. In the upper angle of the wound, a nervous filament was loſt. The fleſhy part of the ſterno-maſtoid, ſterno-hyoid, firſt ſcalenus, coracohyoid, longus colli, of the right ſide, were, in ſome places, of a deep red, as if from blood effuſed; and the longus colli, but no other correſponding muſcle of the left ſide, was affected in a ſimilar manner: the lower extremities of both ſtylo-hyoids, and of both hyogloſſi, were of the deep red colour; the other muſcles employed in diglutition, alſo the tongue, palatum, molle, pharynx, part of the œſophagus, and larynx, upon examination, appeared ſound. There were ſeveral red ſpots on the external ſurface of the tendinous aponeuroſis covering the abdomen, and on many parts of both recti abdominis, there was the ſame deep red colour, which penetrated quite through the ſubſtance of each muſcle; the poſterior ſurface of each being ſtill more generally of that colour, and, in many places, particularly below the navel, the fibres were conſumed, for about an inch of their length, by ulcers containing a reddiſh matter; the poſterior parts of the ſheaths of those muſcles, were, in ſome places, oppoſite to the ulcerations in the muſcles, of a light red. The maſſeters, the temporals, the muſcles which pull the head back, and the viſcera of both great cavities, appeared quite ſound.

CHAP. II.

A Defcription of the Symptoms of Difeafes of the Head, Nerves and Mufcles, taken from thofe Cafes where the Patients recovered, or where the Author had no Opportunity of examining the Bodies after Death.

§ 1. *Lofs of Feeling and of Motion.*

THE total lofs of feeling and of motion in different parts of the body moft frequently happens during a fit of apoplexy, or general infenfibility; fometimes it occurs before fuch fit, and fometimes there is a gradual diminution of feeling with feeblenefs, terminating in complete numbnefs and lofs of motion, without any fit either preceding or accompanying it; the fenfes, memory, and fpeech remaining entire, or to a certain degree impaired. The parts affected are oft contracted fo that the fingers are bent into the hands, and if at any time they are extended by force, which cannot always be done, they return immediately to their former fituation; the leg, alfo, of the fide affected is drawn back towards the thigh, and the foot towards the back of the leg; the paralytic mufcles are at times fuddenly convulfed, which occafions confiderable pain, and they frequently tremble or fhake, and are generally cold. Sometimes the whole body is affected by this diforder, although the one fide more than the other; fometimes the lower half of the body, or only the lower extremities; or the feet and hands only, but moft commonly the whole of one fide, the other remaining unaffected. The ftools, and, at times, the urine, are retained in the beginning, but towards the end, efpecially in bad cafes, both of them run off involuntarily. The pulfe is quick and fmall, ſmall, ſometimes having a kind of vibration, or a very feeble heat is interpoſed between two common pulſations, and there are inſtances, even in favourable caſes, where the pulſe cannot be felt. Amendment, or recovery, is commonly preceded by a painful ſenſation of pricking or ſhooting in the part, after which, in a little time, the feeling and power of motion return, though in a ſlight degree; afterwards, upon thoſe pains going off, the parts recover perfectly their feeling and power of motion, and laſtly their ſtrength.

The recovery of the patient is alſo ſometimes preceded by an eruption of very painful ſpots, raiſed a little above the ſkin. The ſenſe of feeling and motion commonly return firſt in thoſe parts which are neareſt the brain, proceeding gradually downwards from one member to another; but I have known recovery take place in a reverſe order. The diſeaſe continues from one to three months, though, commonly, much longer; ſometimes it proves fatal in five or ſix months from the firſt attack. Sometimes between the paroxyſms the patient is for a year or two ſubject to giddineſs, trembling, *&c.* The cauſe of this diſeaſe is often unknown. Sometimes it happens during a fever, and often ſupervenes ſlight injuries of the head, when, without any fracture of the ſkull, they have been followed by inſenſibility, either immediately or a day or two after the accident, and in all thoſe caſes, the ſide of the body affected is oppoſite to the ſide of the head where the injury has been received.

Do not the pains which commonly precede recovery, point out the uſe of irritating and painful applications to the part?

Does not the eruption of painful ſpots indicate, in a particular manner, the utility of bliſters?

§ 2. *The*

§ 2. *Loſs of Motion with Relaxation of the Parts.*

All or only ſome of the limbs ſuddenly loſe the power of motion, but without any remarkable diminution of feeling; this ſometimes happens without pain; at other times it is accompanied or preceded by very violent pain. When unattended with pain the cauſe is commonly unknown, and, if the feeling is unimpaired, the patient, without any diſagreeable ſenſation, recovers a little ſtrength at firſt, and, finally, the perfect uſe of his limbs.

The caſes attended or preceded by violent pains, moſt frequently occur amongſt people who are employed in the ſeveral trades in which lead or quickſilver are uſed, ſuch as glaziers, plumbers, colour-makers and gilders; there is this difference however, that thoſe perſons who have been expoſed to the fumes of quickſilver, have, beſides the other ſymptoms, almoſt perpetual tremors of the limbs.

In thoſe caſes attended with pain I have alſo obſerved, that the ſuperior extremities are more frequently affected than the inferior, and that the muſcles of the hands are remarkably waſted. The dry belly-ach, formerly deſcribed, commonly, and eſpecially in painters who make uſe of turpentine, precedes or accompanies any affections of the limbs.

The patient, even when he quits his buſineſs, recovers his health but ſlowly; firſt acquiring the power of bending, afterwards of extending the limbs; but there are few inſtances of a complete recovery till after many months, or even years, and a return of the complaint

is

is the almoſt certain conſequence of returning to their former manner of life. The moſt effectual preſervatives are keeping clean, and avoiding, as much as poſſible, all immediate contact with the metal, its calx or fumes.

§ 3. *Loſs of Motion with Contraction of the Parts.**

A woman, aged thirty-one, fell down ſuddenly whilſt walking. She retained her ſenſes, but had violent pains and contractions in the muſcles of both arms, which were ſo ſtrongly bent that all efforts to extend them were ineffectual, and the attempting it only cauſed more exquiſite pain; but though the upper extremities were ſo greatly, the lower were little affected, and ſhe only complained of a ſlight pain in one ancle. After ten hours there was no change in her ſituation; after twenty the pains had ſomewhat abated, and ſhe could move her arms a little. After thirty hours the pains had entirely ceaſed, and ſhe could move all her joints eaſily. In a few days more ſhe recovered the ſtrength of her arms, and in a week, except that ſhe was ſomewhat low-ſpirited, had no farther complaint.

A woman, aged twenty-ſeven, was for ſeveral years ſubject, commonly in cold weather, to fits of coldneſs and of pain in the external parts of the head, face and neck, and in the muſcles within the arch of the lower jaw; during theſe fits ſhe could not bear the ſlighteſt preſſure on the parts affected, and the lower jaw was immoveably fixed, at firſt, almoſt cloſe to the upper one, but afterwards, as the pains diminiſhed, it relaxed ſo far as to admit the point of

* Under this head I have given three caſes, the examples not being ſufficiently numerous to enable me to draw up any general hiſtory of the complaint.

of the finger between the teeth. In about a month or ſix weeks the pains ceaſed, and ſhe perfectly recovered the uſe of the jaw.

A woman, aged twenty-one, whoſe hiſtory I have in part formerly related*, three weeks after receiving the blow on her cheſt, was ſeiſed with a fit, which returned ſometimes every day, though more commonly after an interval of a week, a fortnight, or even a month. When ſhe perceived the approach of the fit, which was preceded by partial muſcular contractions, what ſhe called twitchings or catchings, ſhe laid herſelf on her back in bed, her limbs were immediately ſtretched out, her fingers and toes ſtrongly drawn in, and her head ſo much drawn backwards that her face was turned directly to the head of the bed; in this ſtate ſhe remained ſtruggling for a conſiderable time, her body bent upwards, whilſt the crown of her head was forcibly preſſed againſt the bed, her neck and breaſt were alſo ſwelled, and her belly was repeatedly raiſed forwards; at this time the muſcles every where felt rigid; ſometimes the head, from the poſture above deſcribed, was drawn ſlowly forwards alſo, from ſide to ſide. The eyes were fixed, and not affected by the neareſt approach of objects. She frothed at the mouth and frequently bit her tongue, (which was puſhed out) from the convulſive contractions of the muſcles of the lower jaw. Sometimes ſhe would ſing, or make a noiſe like the barking of a dog, at other times ſhe uttered the moſt doleful cries, after which the muſcles were always ſoft, the limbs relaxed, and her hands opened; in this ſituation ſhe uſed to remain from one to eight hours, and after each fit complained of exceſſive wearineſs and pains all over her body. Sometimes her fits were much ſlighter, ſome of her limbs only being contracted; and, though ſhe loſt her ſight, retaining her other ſenſes. During the intervals of the

M fits,

* Vide page 45.

fits, ſhe was troubled with twitchings and tremblings, either of her whole body, of one ſide, or of a particular limb only, and theſe were always greatly increaſed from any fright or flurry. She frequently gnaſhed her teeth, and ſometimes with ſo much violence when ſhe was drinking, as to break the cup. She alſo complained of head-ach, giddineſs, dimneſs of ſight, lowneſs of ſpirits, coldneſs of her lower extremities, and, ſometimes, had cold ſweats; her pulſe was between eighty and ninety. Her fits remained violent near fourteen months, but became more moderate after the burſting forth of matter from her ſide, and did not afterwards affect her ſenſes, and in four months after this event took place, they, as well as the concomitant ſymptoms, diſappeared entirely, and have not ſince, now near four years, in any degree returned.

§ 4. *Perpetual involuntary Motion.*

Moſt commonly after a fright, ſometimes after convulſions, hyſterical or fainting fits, and ſometimes nothing remarkable having preceded, the patients are ſeized with a perpetual involuntary motion, but without pain, either of all the limbs and ſpine, or of both arms, though unequally, or of the arm and leg of one ſide, or of one arm only, or of the belly and breaſt, which laſt motion is much quicker than that of reſpiration. Theſe motions are ſometimes ſo violent that, when general, the patients cannot lie in bed, and when one arm only is affected, its motion will throw them down, if while walking they happen to be off their guard; by the perpetual rubbing the cuticle is ſometimes abraded from the inſides of the fingers. Theſe motions ſometimes encreaſe in violence in the evening, and on alternate days, and, when going off intermit in the forenoon. The patients

patients ſometimes laugh or cry, are troubled with a hiccough, or ſmack with the tongue and lips; the tongue is often puſhed out very far, and the lower jaw is in perpetual motion; if they happen, as is ſometimes the caſe, to ſlumber for a little, the parts, during ſleep, are at reſt. Sometimes they complain of pain in the throat, breaſt and neck. The ſpeech is commonly affected, but the ſenſes are entire. Both ſtools and urine are retained longer than is natural. The pulſe is ſmall and ſometimes quick. By a ſuperior external force the motion of a limb may be ſtopt for a little, commonly without any inconvenience to the patient; but in one caſe, when the arm was held, the patient ſunk quietly into a fit, as if ſhe had been aſleep, the other limbs retaining the poſture they happened to be in when the fit began, and when, on letting go the arm, the motions of it returned, the feeling and the power of motion in the other limbs returned alſo; ſometimes the motions of the arms, of their own accord, alternately ceaſed and returned, and the ſame conſequences followed. The pulſe did not vary in the different ſtates. In the ſame patient the left arm was, without any inconvenience, always at reſt when ſhe lay down, and always in motion when ſhe ſat upright or ſtood, the right arm was not affected by change of poſture, till her recovery was advanced, and then it was influenced by it in the ſame manner as the left arm had formerly been.

The ſubjects of this diſorder are women and children. The duration is commonly not longer than one or two months; but ſometimes the motions of particular limbs continue for ſeveral years, and in one caſe, where they had been violent, the limb inflamed. Patients are apt to ſuffer a relapſe, eſpecially when the diſeaſe originated in a fright.

Does the relief which is afforded during ſleep, direct us to the uſe of opiates?

C H A P. III.

Obſervations on the Effects of Remedies in Diſeaſes of the Head, &c.

THE ſenſe of feeling and the power of motion were commonly encreaſed after the application of bliſters to the nape of the neck, or to the arms, when theſe were the parts affected, and to the os ſacrum, when the lower limbs were affected. Liniment. ſapon: rubbed on the parts appeared to have a ſimilar effect—After receiving once in three or four days about a dozen of ſlight ſhocks of electricity, the ſenſe of feeling was in a few hours, and the power of motion in a few days encreaſed—After going into a warm bath the pulſe roſe, and the ſtrength of the limbs was encreaſed a little: Painters, whoſe wriſts were weak, found ſome benefit after having repeatedly held their hands and the lower parts of their arms in the warm moiſt grains of malt. After hot medicines, ſuch as ſal. C. C. vol. muſtard-ſeed, horſe-raddiſh-root, gum gauiacum, ſaffron, and ſome other ſudorifics, a glowing was felt in the affected parts, and was followed by ſweating, and, in ſome caſes, when bliſters alſo had been applied, the ſenſe of feeling and power of motion were perfectly reſtored. After Peruvian bark and ſteel medicines, tremors and weakneſs of the limbs were diminiſhed. In regard to the perpetual involuntary motions, the moſt remarkable relief, or rather almoſt perfect recovery, happened to a woman, who, having a ſecond time had

this

this diſorder conſtantly in her arms for three years, took muſk a little longer than a fortnight, to the quantity firſt of one drachm, and afterwards to that of one drachm and an half each day; ſhe ſweated a little during the courſe, and was giddy from the encreaſed doſe. Opiates and fetid gums, with ſalt of hartſhorn, ſeemed alſo to have in theſe caſes very good effects.

As the diſeaſes hitherto deſcribed are principally ſuch as ariſe from the affections of particular organs, I have been fuller in giving anatomical deſcriptions than may perhaps be neceſſary hereafter. And therefore at the cloſing of this part of my undertaking, I reckon it incumbent upon me to ſay ſomething of the advantages, which, may be derived from the diſſection of morbid bodies.

And here it firſt occurs, that it muſt ſurely give a Phyſician great ſatisfaction and pleaſure to find, by the appearances, that he has underſtood a diſorder and treated it properly; but this being a kind of delicate luxury in ſcience, reliſhed only by the moſt liberal minds, and therefore a ſuperfluity, we muſt next enquire, whether this modern method of arriving at knowledge, may not be attended with ſome more ſolid advantages, advantages really conducive to the health and happineſs of mankind.

Though there are many diſeaſes which have not hitherto been in any degree illuſtrated by diſſections, yet the great light which has been thrown upon others, by the accidental diſſections of anatomiſts, ignorant for the moſt part of the complaints which preceded death, or who learned them only by hearſay, and after the diſſections had been performed, is a ſufficient earneſt of the great encreaſe in the knowledge of diſeaſes which might be made, were Phyſicians, who have known the complaints, to examine more minutely

minutely and attentively than they commonly do, the bodies of the dead. A perſon who has carefully performed, or even attended to the diſſection of one caſe, will afterwards look upon ſimilar caſes with a more piercing eye than before; for as ſymptoms ſuggeſt to the mind the ideas of certain changes in the body, ſo, on the other hand, the obſervation of certain changes in the body ſuggeſts the ideas of certain ſymptoms connected with them, which, though neceſſary to a full knowledge of the diſeaſe, would otherwiſe eſcape the notice of the Phyſician.

Diſſections have alſo led to ſeveral uſeful and neceſſary methods of treating diſeaſes, which were formerly unknown, and have, likewiſe, ſhewn the inutility and impropriety of many common methods of practice. They tend more than any thing whatever to ſhew the inſignificancy and the futility of many highly and long applauded remedies, and thus the mind being ſet at liberty from a ſlaviſh implicit faith in their efficacy, its powers may be directed to more worthy objects.

If, therefore, by diſſections, in the performing of which, diſeaſes were frequently only a ſecondary object, ſo great a progreſs has been made in the knowledge, and in the treatment of them, let thoſe who wiſh to promote the great intereſt of mankind, avoid joining themſelves to the lazy tribe of deſpondents, who aſſert that Phyſick cannot be improved, and who, on that pretence, give way to their innate love of indolence and ſloth. Let the generous few rather hope, nay, let them be aſſured, that by their united and continued efforts, the knowledge of diſeaſes may be very highly and eſſentially promoted.

EXPERI-

EXPERIMENTS

ON

VARIOUS SUBJECTS.

EXPERIMENTS,

DIETETICAL AND STATICAL.

EXPERIMENTS

ON

DIET.

INTRODUCTION.

ALTHOUGH air is more immediately neceſſary to life than food, the knowledge of the latter ſeems of more importance; it admits certainly of greater variety, and a choice is more frequently in our power. A very ſpare and ſimple diet has commonly been recommended as moſt conducive to health, but it would be more beneficial to mankind if we could ſhew them that a pleaſant and varied diet was equally conſiſtent with health as the very ſtrict regimen of Cornaro, or the Miller of Eſſex. Theſe and other abſtemious people, who, having experienced the greateſt extremities of bad health, were driven to temperance as their laſt reſource, may run out in praiſes of a ſimple diet, but the probability is, that nothing but the dread of former ſufferings, could have given them reſolution to perſevere in ſo ſtrict a courſe of abſtinence; which, perſons who are in health, and have no ſuch apprehenſion, could not be induced to undertake, or, if they did, would not long continue.

In all caſes great allowance muſt be made for the weakneſs of human nature; the deſires and appetites of mankind muſt, to a certain degree, be gratified, and the man, that wiſhes to be moſt uſeful, will imitate the indulgent parent who, whilſt he endeavours to promote the true intereſts of his children, allows them the full enjoyment of all thoſe innocent pleaſures which they take delight in. If poſſibly it could be pointed out to mankind that ſome articles uſed as food were hurtful, whilſt others were in their nature innocent, and that the latter were numerous, various and pleaſant, they might, perhaps, from a regard to their health, be induced to forego thoſe which were hurtful, and confine themſelves to thoſe which were innocent. To eſtabliſh ſuch a diſtinction as this, from experiment and obſervation, is the chief object of my enquiry: and I confeſs it will afford me a ſingular pleaſure if I can prove, by experiment, that a pleaſant and varied diet is equally conducive to health, with a more ſtrict and ſimple one; at the ſame time I ſhall endeavour to keep my mind unbiaſſed in my ſearch after truth, and, if a ſimple diet ſeems the moſt healthy, I ſhall not heſitate to declare it.

But before entering upon the preſent, or any other inveſtigation, it may not be improper to attend to a diſtinction, made by my Lord Bacon, between uſeful and curious knowledge; the latter, indeed, or Experimenta lucifera, he recommends, as nearly of equal importance with the former, or Experimenta fructifera, though to me they appear widely different. The only teſt of the utility of knowledge is, its promoting the happineſs of mankind; which, though

though the Experimenta lucifera may do at ſome future period, the Experimenta fructifera, as having directly and immediately this effect, are ſurely to be preferred; and, therefore, though I admit that all knowledge is deſireable, from the pleaſure it affords, yet, conſidering the ſhortneſs of human life, and the very narrow limits of human abilities, and conſidering alſo that there are many things ſtill unknown which might be of advantage to ſociety, it may be doubted whether every perſon be not in ſtrict duty bound to direct his whole attention to the cultivation of uſeful knowledge.

In the courſe of the preſent enquiry, I have ſometimes doubted whether an accurate attention to the diſcharges of the body be not more a matter of curioſity than of uſe; and, if our attention ſhould not be chiefly directed to obſerve the different effects of food on the body; whether, for inſtance, it agrees or diſagrees with the ſtomach, is more or leſs nouriſhing, has the quality of invigorating, or of occaſioning lazineſs and inactivity, if it enlivens or deadens the faculties, and if it creates or allays the ſeveral appetites and deſires.

If after what I have ſaid, I ſhall be thought to have indulged myſelf with attending too accurately to the diſcharges of the body, it muſt be partly imputed to my deſire to avoid the appearance of entertaining a doubt, that what ſome of the firſt names in Phyſic have thought deſerving their attention were things of little moment.

Dr. Stark, before he began his Experiments on Diet, had collected some facts on the subject, and had made some observations relative to digestion, which I have introduced in this place, imagining that they would not be unacceptable to the public.

Facts relative to Diet.

Dr. B. Franklin, of Philadelphia, informed me, that he himſelf, when a journeyman Printer, lived a fortnight on bread and water, at the rate of 10 ℔ of bread *per* week, and that he found himſelf ſtout and hearty with this diet.

He likewiſe told me, that he knew a Gentleman, who, having been taken by the Barbary Corſairs, was employed to work in the quarries, and that the only food allowed him was barley, a certain quantity of which was put into his pockets every morning; water he found at the place of labour; his practice was, to eat a little now and then, whilſt at work, and, having remained many years in ſlavery, he had acquired ſo far the habit of eating frequently and little at a time, that when he returned home his only food was gingerbread-nuts, which he carried in his pocket, and of which he eat from time to time.

By Sir John Pringle I was told, that the inhabitants of Zephalonia, during ſome parts of the year, live wholly on currants. He alſo

alſo ſaid, that he knew a Lady, now ninety years of age, who eat only the pure fat of meat.

I learned from Dr. Mackenzie, that many of the poor people near Inverneſs, never took any kind of animal food, not even eggs, cheeſe, butter or milk.

Mr. Hewſon informed me, that Mr. Orred, a Surgeon at Cheſter, knew a ſhip's crew, who being detained at Sea after all their proviſions were conſumed, lived, one part of them on tobacco, the other on ſugar: and that the latter generally died of the ſcurvy, whilſt the former remained free from this diſeaſe, or ſoon recovered.

Dr. Cirelli ſays, that the Neapolitan Phyſicians frequently allow their patients, in fevers, nothing but water for forty days together.

Mr. Slingſby has lived many years on bread, milk and vegetables, without animal food or wine: he has excellent ſpirits, is very vigorous, and has been free from the gout ever ſince he began this regimen.

Dr. Knight has lived alſo many years on a diet ſtrictly vegetable, excepting eggs in puddings, milk with his tea and chocolate, and butter—He finds wine neceſſary to him—Since he lived in this manner he has been free from the gout.

Obſervations

Obſervations on Digeſtion.

A woman, who was in the practice of dram-drinking, after taking an emetic, vomited many pieces of fat, ſome pieces of griſtle, and only one very ſmall bit of the lean of veal, which ſhe had eat twenty-four hours before. She brought up, likewiſe, a bit of apple, and ſome pieces of the ſkins of roaſted apples which ſhe had eat twenty hours before. Alſo part of a brown cruſt of bread which ſhe had eat about three hours before.

A young man, ill of a fever, having taken an emetic, vomited ſome fat broth, with bits of bread, which he had eat three hours before.

A young man, who had been ſlightly indiſpoſed about a week, after taking an emetic, brought up ſome mutton, which he had eat three hours before, and nearly in the ſame ſtate in which he eat it.

A girl, in a fever, vomited ſpontaneouſly, and brought off her ſtomach ſome fiſh, which ſhe had eat three hours before.

A man, with purging, head-ach, *&c.* brought off his ſtomach, by an emetic, ſome very diſagreeable bitter ſtuff, but without any appearance of bread and butter, of which he had eaten very heartily about ſeven hours before.

A girl, ſubject to fits, after taking an emetic, vomited ſome oyſters, which ſhe had eat three hours before, but there was no appearance of veal, which ſhe had eat twenty-ſeven hours before.

A girl,

A girl, ſubject to pains in her ſtomach, after taking an emetic, vomited many pieces of the ſkin, but ſcarcely any of the lean of roaſted veal, which ſhe had eaten ſix hours before.

A woman, with a tertian fever, head-ach, *&c.* after taking an emetic, brought up ſome mutton, very little altered. She had not eat it long before.

Mrs. I—m informed me, that her ſon, a little boy, and her daughter, a delicate girl, vomited, one morning, ſome beef, which they had eat at dinner, between three and four o'clock the preceding day.

ABSTRACT *of a* JOURNAL *kept during a Courſe of Experiments on Diet.*

N. B. The weight of the ſolid food and ſtools, is marked in Troy weight, that of the body in Avoirdupois; the quantity of liquids was determined by wine meaſure—The weight of my body, dreſſed in my uſual clothes, at the beginning of theſe Experiments was, 12ſt. 3lb. or 171lb. Avoirdupois.

EXPERIMENT I.

Diet of Bread and Water.

	State of the atmoſphere during the period.	Daily allowance of food	Daily loſs of weight.	Number and total weight of ſtools.	Weight of my body at the end of the period.
Firſt period, from the 12th to the 24th of July, 1769.	o	Bread 20 oz. Water 4 lb.	5 oz. 5 dr.	5 ſtools, weighing 7 oz. 5 dr.	11 ſt. 12 lb. 8 oz.
Second Period, from the 24th of June to the 13th of July.	Thermometer from 60 to 70. Weather commonly ſerene, ſometimes cloudy, ſeldom rain.	Bread, 30 oz. Water, 2 lb.	6 oz. 10 dr.	7 ſtools, weighing 10 oz. 5 dr.	11 ſt. 10 lb. 8 oz.

Third

	State of the atmosphere during the period.	Daily allowance of food.	Daily loſs of weight.	Number and total weight of ſtools.	Weight of my body at the end of the period.
Third period, from the 13th to the 19th of July.	Thermometer from 60 to 73. Often ſerene, ſometimes cloudy.	Bread, 30 oz. Water, 2 lb.	6 oz. 10 dr.	1 ſtool, weighing 2 oz. 5 dr.	11 ſt. 7 lb. 8 oz.
			Daily gain in weight.		
Fourth period, from the 19th to the 26th of July.	Thermometer from 63 to 66. Commonly cloudy, ſometimes rain, ſometimes ſerene.	Bread, 38 oz. Water, 3 lb. 8 oz.	3 oz. 6 dr.	3 ſtools, weighing 2 lb. 1 oz. 3 dr.	11 ſt. 9 oz. 8 dr.

REMARKS.

To determine how long the food is uſually retained in the body, I repeatedly ſwallowed muſtard or carraway-ſeeds and obſerved, that when coſtive, they did not paſs with the firſt, but with the ſecond and third ſtool; and, commonly, after thirty-ſix or forty-eight hours; when open in the body, they came away with the firſt ſtool, the next morning.

Before I began regularly this Courſe of Experiments, I had, for ſeveral weeks been, now and then, making trial of it; ſometimes, inſtead of water, I took, in the morning, a weak infuſion of tea, aſſafras, or of ſome herb, but without milk or ſugar. My ſtools were of a ſmooth conſiſtence and ſlimy ſurface, like clay.

Although upon the allowance of twenty ounces of bread, I was hearty, in good ſpirits, and had ſome deſires, yet I found it neceſſary to encreaſe it, not only as I fell away, but becauſe I was often very hungry.

On the allowance of thirty ounces, I ſometimes, immediately after eating, had a little wind upwards, and ſometimes, though rarely, a little downwards. My ſtools were gradually ſofter. I ſtill fell away very viſibly; had hardly any deſires, though hearty in other reſpects. Sometimes I felt a ſlight ſickiſhneſs and want of appetite, which went off after eating a bit of bread.

Imagining that the ſickneſs might be owing to my taking an over proportion of liquid, I endeavoured, during the third period, to aſcertain how much liquid was abſolutely neceſſary to the quantity of bread I eat, and found, that though I could eaſily eat my common meal of ten ounces, without any liquid, and was not at all thirſty, even for ſome time after, yet in two or three hours, an intolerable thirſt came on, which could not be allayed by leſs than ten ounces of liquid.

I likewiſe found that when I drank leſs than two pints a day I was thirſty in the evening, and had a ſlight pain in my ſtomach. formerly I uſed to make eleven or twelve ounces of urine at a time, but now five or ſix ounces brought on the inclination, and my water was high coloured During the third period I was one day irregular, having ate about four ounces of meat, and drank two or three glaſſes of wine. At the concluſion of it,

I was

I was perfectly hearty, my head clear, often hungry, but never had any desires.

When I eat thirty-eight ounces of bread (the allowance during the fourth period) at five or six times, my appetite was not more than satisfied, but if I made fewer meals I found my appetite satiated.

I sometimes varied my daily quantity of bread, by taking from the allowance of one day and adding to that of the day following, but I found that the most I could eat in one day was forty-six ounces, and that the greatest quantity I could eat at one time, without uneasiness, was twenty ounces; that the sensation of hunger began four hours after eating this quantity, when I could eat twenty ounces more. I once forced myself, to eat, at one meal, in an hour and ten minutes, thirty ounces of bread, I brought up some wind off my stomach whilst I was eating it, had afterwards much noise in my bowels, and in a few hours a bolar stool, weighing one pound; I continued uneasy during the whole of the evening, but was quite well and hungry next morning. During this last period I sometimes had desires (Venus bis) but never before, since I began this Course of Experiments.

By Experiment, I determined, the quantity of saliva secreted in half an hour, to be whilst the parts were at rest, four drachms, whilst eating, five ounces four drachms.

EXPERIMENT II.

Diet of Bread and Water with Sugar.

	State of the atmofphere during the period.	Daily allowance of food	Daily gain of weight.	Number and total weight of ftools.	Weight of my body at the end of the period.
First period, from the 26th of July, to the 3d of Auguft.	Thermometer from 62 to 66. Weather commonly cloudy.	Bread 34 oz. Sugar 4 oz. Water $3\frac{1}{2}$ lb.	2 oz.	Purging.	11 ft. 10 lb. 8 oz.
Second period, from the 3d to the 9th of Auguft.	Thermometer from 64 to 74. Weather commonly ferene.	Bread, 30 oz. Sugar 8 oz. Water, $3\frac{1}{2}$ lb.	Weight of the body ftationary.	2 ftools, weighing 10 oz. 4 dr.	11 ft. 10 lb. 8 oz.
			Daily lofs of weight		
Third period, from the 9th to the 14th of Auguft.	Thermometer from 63 to 66. Weather ferene, fometimes rain.	Diet irregular.	1 lb.	Purging.	11 ft. 6 lb.

Fourth

Period	State of the atmosphere during the period.	Daily allowance of food.	Daily gain of weight.	Number and total weight of ftools.	Weight of my body at the end of the period.
Fourth period, from the 14th to the 19th of Auguft.	Thermometer from 61 to 63. Serene weather, fometimes rain.	Bread 26 oz. Water 2 lb. 5 oz.	Nearly 3 oz,	Purging.	11ft. 7lb.
Fifth period, from the 19th to the 24th of Auguft.	Thermometer from 59 to 61. Weather ferene, fometimes rain.	Diet irregular.	Weight of the body ftationary.	1 loofe ftool.	11ft. 7lb.

REMARKS.

Sugar feemed to increafe the flow of faliva into the mouth, for with fugar I could eat more bread at a time, than I could poffibly do without this addition.

In the afternoon of the firft day after ufing fugar I paffed a good deal of fetid wind downwards, and early next morning had a liquid ftool. I had afterwards three loofe ftools, weighing one pound five ounces; weak defires, (Venus femel) during the firft period.

After I began to ufe fugar with my bread, I found that a fmaller quantity of liquid prevented thirft than when I eat bread alone.

alone. With my preſent diet, of thirty ounces of bread, and eight ounces of ſugar, two pints of liquid a day are ſufficient to allay my thirſt; whereas, when I ate thirty-eight ounces of bread, without ſugar, I found that three pints and a half of liquid were abſolutely neceſſary.

I commonly ate eight ounces of ſugar at a meal, without any inconvenience, and became hungry three hours after it; my appetite was not at all cloyed with the ſugar. I paſſed hardly any wind either way, and never had any deſires.

On the 10th I ate, at three different times, before one o'clock, twenty ounces of ſugar, and, though I ate the laſt of it with reluctance, and was ſickiſh after it, yet it did not ſatisfy my appetite. At two o'clock I became very hungry, and at three began to eat bread with great pleaſure, and ate twenty ounces of it, drinking two pints and a half of water, which I found ſufficient to allay my thirſt.

On the 11th I ate twenty-four ounces of bread, and ſixteen ounces of ſugar, but the laſt part of it with great abhorrence. I now perceived ſmall ulcers on the inſide of my cheeks, particularly near a bad tooth, in the lower jaw, of the right ſide; the gums of the upper jaw, of the ſame ſide, were ſwelled and red, and bled when preſſed with the finger, the right noſtril was alſo internally red or purple, and very painful. I had one thin ſtool.

On the 12th I ate thirty ounces of bread, with ten of ſugar, had little appetite for ſupper, and after it a thin ſtool.

On the 13th, having been extremely ill, during the night, with pains in my bowels and ſweating; at day-break, I had a large thin ſtool, and two liquid ſtools afterwards, but paſſed no wind, nor was troubled with any in my bowels. I had no appetite for breakfaſt, could not taſte ſugar, dined on a few ounces of meat, with about twelve ounces of bread, and drank two or three glaſſes of wine.

On the 14th I perceived ſeveral ſmall purple ſtreaks on my right ſhoulder, but the ſores in the inſide of my mouth, and my gums were better, and my noſtril leſs painful.

On the 15th the affection of my gums, though leſs in degree, had become more general, having ſpread to the left ſide, their ſemilunar edges were of a deep red, and ſeveral drops of blood iſſued from my right noſtril.

N. B. Until the 18th I had, every day, three or four liquid ſtools, containing ſome clear gelatinous ſubſtance, and felt but little pain or wind in my bowels—on the 18th and 19th I had one ſtool each day.

On the 18th, ſome part of the gums of both jaws, and on both ſides, were red and ſwelled, but none of them of that purple colour, nor ſo apt to bleed as ſome days ago, the ſores in my mouth were healing, and the ſtreaks on my ſhoulder almoſt gone. I never had the ſmalleſt deſires.

From the 19th to the 24th, my food was thirty ounces of bread with three pints of water every day, excepting on the 22d, when I dined

I dined heartily on meat and fruit, and drank ſome wine. Venus ſemel.

N. B. On the 21ſt I made an experiment with two drachms of fæces, of a pilular conſiſtence, which I had voided, after having lived about a week on bread and water; they were waſhed four or five times in about ſix ounces of water, which was thereby rendered milky; but after ſtanding ten or twelve days, and depoſiting a ſediment, it became again almoſt tranſparent; the reſiduum, ſaved on the filtring paper, weighed one ſcruple and half a grain, was of a darkiſh green colour, and perfectly inodorous—Bread, treated in a ſimilar manner, occaſioned no milkineſs, and the water, inſtead of becoming putrid, was converted into a weak vinous liquor.

EXPERIMENT

EXPERIMENT III.

Diet of Bread and Water with Oil of Olives.

	State of the atmosphere during the period.	Daily allowance of food	Daily gain or loss of weight.	Number and total weight of stools.	Weight of my body at the end of the period.
First period, from the 24th to the 30th of August.	Thermometer from 59 to 62. Weather serene, sometimes rainy.	Bread 30 oz. Oil of Olives $2\frac{1}{2}$ oz Water 3 lb.	Gained nearly 5 oz. 3 dr.	2 stools, weighing 1 lb. 4 oz. 6 dr.	11 st. 9 lb.
Second period, from 30th of August to the 5th of September.	Thermometer from 63 or 64. Weather serene or cloudy.	Bread, 30 oz. Water, 3 lb.	Lost nearly 9 oz. 3 dr.	1 stool, weighing 4 oz. 4 dr.	11 st. 5 lb. 8 oz.
Third period, from the 5th to the 13th of September.	Thermometer from 57 to 66. Weather commonly rainy.	Diet irregular.	0	Purging.	11 st. 13 lb. 8 oz.

 REMARKS

REMARKS.

Two ounces of oil, taken at one meal, was ſo large a quantity as to be diſagreeable; three ounces in the day occaſioned ſome uneaſineſs in my bowels; and four ounces griped me very much—I had now and then ſome wind upwards, and ſometimes, after being a little griped, paſſed ſome downwards; my appetite was ſufficiently ſatisfied; I was ſometimes a little thirſty, and frequently had deſires in the night.

On the 23d of Auguſt, a large double tooth, which had been very troubleſome to me, during the time, and even after the ſugar diet, was extracted from the lower jaw; the day following I had great pain in the part from whence the tooth was taken, and the gum appeared ſomewhat black.

The ſecond night I had no ſleep from the exceſſive pain, and an abominably putrid ſlough was formed. The gums in the neighbourhood of the ſore ſwelled more than ever and became in part livid, with a fetid white ſtuff round their edges, whilſt the gums immediately over the ſore were black and inſenſible. My appetite was keen, notwithſtanding this complaint in my mouth, and was not ſatisfied until the 5th of September, when I loſt it entirely, and became dull, I never had any wind in my ſtomach, and ſeldom in my bowels. No deſires. I commonly kept ſome powder of bark on the ſore, and waſhed it frequently with diluted vinegar.

On

On the 5th of September the flough was fmoother, not fo fetid or difagreeable, but the affection of the gums was more general, and fome of them a little eroded.

On the 6th I had a loofe ftool in the morning, little or no appetite.

On the 7th, ftill no appetite, I had five loofe ftools, with griping and wind, and the ftools partly confifted of a kind of gelatinous matter. On going to bed in the evening I was feized with coldnefs and fhivering, had fourteen watery ftools in the night, with great pain and wind in my bowels, &c.

On the 8th I was fo weak and low that I almoft fainted in walking acrofs my room; had four or five loofe ftools in the courfe of the day, was fick, and my tongue foul. Having taken fifteen grains of ipecacuanha, I vomited, firft a clear liquor, of the colour of Burgundy, afterwards a brown and extremely bitter liquor. In the evening I obferved that the flough on the fore, and fome parts of the gums had become black, whilft the gums of the upper jaw, oppofite the fore, were fwelled, fo as almoft to reach the extremity of the eye-tooth; and I fpat, in confiderable quantity, a very difagreeable, fetid, yellowifh fluid. I took half an ounce of the extract of the bark, and had three ftools, but they were not fo thin as before.

On the 9th, although I was much better, my pulfe was ftill very low, and I was apt to faint whenever I got out of bed; fome black floughs were feparated from the gums, which now put on a more favourable appearance. The eminences or papillæ, which to me are natural on the infide of my legs and thighs, were red or purple, and the diſcolouration of the ſkin ſpread beyond the eminencies; there were alſo a few light brown ſpots on ſeveral parts of my lower extremities. I took an ounce and an half of the extract of bark, with ſome mulled Port wine, which had no very ſenſible effect; but I found myſelf greatly revived by a baſon of mutton broth, which was almoſt the firſt food I had taken ſince the 5th; I had two ſoft ſtools. I continued to take the bark for a few days longer, and lived freely on animal food, milk and wine, until the 18th; when I felt myſelf quite recovered.

EXPERIMENT IV.

Diet of Bread and Water, with Milk.

	State of the atmoſphere during the period.	Daily allowance of food.	Daily gain of weight.	Number and total weight of ſtools.	Weight of my body at the end of the period.
Firſt period, from the 18th to the 22d of September.	Thermometer from 57 to 62. Weather ſerene,	Bread, 30 oz. Water, 3 lb. Milk, 4 lb.	2 oz.	4 ſtools, weighing 3 lb. 10 oz.	12 ſt.
Second period, from the 22d to the 26th of September.	Thermometer from 55 to 57. Weather cloudy or ſerene.	Bread, 30 oz. Water, 3 lb. Milk, 4 lb.	2 oz.	2 ſtools, weighing 1 lb. 4 dr.	12 ſt. 8 oz.
Third period, from the 26th to the 29th of September.	Thermometer from 55 to 59. Weather rainy, or ſerene.	Bread 30 oz. Water 3 lb.	Daily loſs of weight. 10 oz, 5 dr.	2 ſtools, weighing 5 oz. 4 dr.	11ſt. 12 lb. 8 oz.
Fourth period, from the 29th of September to the 2d of Oct.	Thermometer 54 or 55. Weather ſerene, cloudy, or rain.	Diet irregular.	———	———	12 ſt.

R E M A R K S.

By the 18th of September the ſore in that part of the gums from which the bad tooth had been extracted, was perfectly healed; and the gums, though ſtill a little ſwelled, were daily getting better. My ſtools were commonly ſoft, and of a buff colour; I was ſometimes a good deal griped. (Venus bis.)

From the 22d to the 26th my ſtools were very hard, forced off with great difficulty and pain, and were covered with blood; I was quite ſtout and hearty, and had, ſometimes, deſires.

On the 29th although the gums were not to appearance worſe, yet I frequently ſucked blood from them, and my finger, after touching them, had an offenſive ſmell; what I ſpit was yellowiſh and fetid. I had obſerved none of theſe ſymptoms before, ſince my ſevere illneſs.

From the 29th of September to the 2d of October, I lived rather highly, on animal food, and, from being coſtive, I became looſe in my body. The bleeding of the gums was leſs perceptible, and they had now no offenſive ſmell. (Venus ſemel.)

EXPERIMENT V.

Diet, Bread and Water, with roasted Goose.

	State of the atmosphere during the period.	Daily allowance of food.	Daily loss of Weight.	Number and total weight of stools.	Weight of my body at the end of the period.
First period, from the 2d to the 6th of October.	Thermometer from 47 to 52. Weather cloudy or rain.	Bread, 30 oz. Roasted Goose, 6 oz. Water, 3 lb.	4 oz.	1 stool, weighing 9 oz. 6 dr.	11 st. 13 oz.
			Daily gain in weight.		
Second period, from the 6th to the 10th of October.	Thermometer ——— Weather commonly serene.	Bread, 30 oz. Roasted Goose, 6 oz. Water 3 lb.	3 oz.	Loose stools.	11 st. 13 lb. 12 oz.
			Daily loss of weight.		
Third period, from the 10th to the 14th of October.	Thermometer about 50. Weather serene.	Bread, 30 oz. Roasted Goose, 6 oz. Water 3 lb.	3 oz.	Loose stools.	11 st. 13 lb.

First

	State of the atmosphere during the period.	Daily allowance of food.	Daily loſs of weight.	Number and total weight of ſtools.	Weight of my body at the end of the period.
Fourth period, from the 14th to the 19th of October.	Thermometer about 50. Weather ſerene.	Irregular.	3 oz.	Looſe ſtools.	12 ſt. 1lb. 4 oz.
			Daily gain of weight.		
Fifth period, from the 19th to the 21ſt of October.	Thermometer 56. Weather cloudy.	Bread, 30 oz. Roaſted Gooſe, 8 oz. Water 3 lb.		2 looſe ſtools.	12 ſt. 1 lb. 8 oz.

R E M A R K S.

I had ſucked blood from my gums till the 3d of October, but none afterwards; the ſwelling of the gums of the upper jaw had ſubſided, and they ſeemed to be quite well, whilſt thoſe of the lower jaw were daily mending; in every reſpect I was hearty and vigorous both in body and mind. (Venus ter.)

On the 7th, I had a head-ach, and little appetite for food. One looſe ſtool.

On the 8th, had two looſe ſtools; my gums were rather worſe, and I brought away a little blood by ſucking them.

Between

Between the 10th and 14th had two liquid ftools; my gums quite well. (Venus bis.)

From the 14th to the 19th, lived freely on animal food.

From the 19th to the 21ft, was fometimes a little thirfty, and my appetite was rather more than fatisfied; violent defires; paffed a good deal of wind downwards. (Venus bis.)

EXPERIMENT VI.

Diet of Bread and Water, with boiled Beef.

	State of the atmosphere during the period.	Daily allowance of food.	Weight of my body stationary.	Number and total weight of stools.	Weight of my body at the end of the period.
First period, from the 21st to the 24th of October.	Thermometer from 47 to 54. Weather serene or cloudy.	Bread, 30 oz. Boiled Beef, 6 oz. Water, 3 lb.		1 stool, weighing 4 oz. 5 dr.	12 st. 1 lb. 8 oz.
Second period, from the 24th to the 28th of October.	———	Bread, 30 oz. Boiled Beef, 4 oz. Water, 3 lb.	———	2 soft stools, weighing 9 oz. 12 dr.	12 st. 8 oz.

REMARKS.

Of the beef, nearly a third part was fat. I found six ounces too much for one meal, and therefore I divided it into two. (Venus bis.)

Upon the allowance of four ounces, I did not find my appetite sufficiently satisfied, although I passed less wind downwards than when I ate the six ounces. I was never in the least heavy or dull after any meal; had no venereal desires, but a keeness for study. I sometimes infused some flowers of lavender, or rosemary, in the water I used, but found nothing so agreeable as green tea.

A REPETITION

A Repetition of EXPERIMENT II.

Diet of Bread and Water with Sugar.

	State of the atmoſphere during the period.	Daily allowance of food	Number and total weight of ſtools.	Loſs of weight at the end of the period.
Firſt period, October 28, 29, 30.	Thermometer from 48 to 52. Weather cloudy, with much rain.	Bread, 30 oz. Loaf Sugar, 6 oz. Water, 3 lb.	2 firm ſtools, weighing 6 oz. 9 dr.	1 lb.
Second period, October 31, November 1.	Thermometer from 53 to 55. Weather rainy.	Bread, 30 oz. Loaf Sugar, 6oz. Water, 3 lb.	1 very firm ſtool, weighing 3oz. 5 dr.	Encreaſe of weight at the end of the period. 1 lb.

REMARKS.

Being now in perfect health, and my gums apparently ſound, I thought it a proper time to aſcertain, by experiment, whether the affection of my gums, and the other complaints with which I had formerly been attacked, were occaſioned by ſugar, or were owing to my having perſevered too long in a diet of bread and water.

On the 28th, I brought up a good deal of wind off my ſtomach, after each meal; on the 29th a little; but on the 30th none. I paſſed ſcarcely any downwards, and what I did paſs, was much leſs fetid than when I lived on beef. My appetite was ſufficiently ſatisfied, and, excepting on the firſt day, that I perceived a little clammineſs in my mouth, I was not in the leaſt thirſty.

During the ſecond period, or the two laſt days, there happened a great irregularity in my weight, for which, not having attended to the quantity of my urine or perſpiration, I can aſſign no reaſon. My weight was encreaſed 1 ℔ the firſt day, and was leſſened 2 ℔ the next. My appetite was hardly ſatisfied; I was never thirſty. I paſſed a little wind downwards, not at all fetid. I had no deſires. My gums were not in the leaſt affected.

A REPETITION

A Repetition of EXPERIMENT VI.

Diet of Bread, with boiled Beef, and Water.

Period. November 2, 3, 4, 5, 6, 7.	State of the atmoſphere during the period.	Daily allowance of food.	Number and total weight of ſtools.	Increaſe of weight at the end of the period.
	Thermometer from 53 to 55. Weather rainy and cloudy, ſeldom ſerene.	Bread, 30 oz. Boiled Beef, ¼ part of which was fat, 6 oz. Water, 3 lb.	4 pretty firm ſtools, weighing 1 lb. 1 oz. 3 dr.	1 lb. 8 oz.

R E M A R K S.

On the firſt day of this period I brought off a little wind from my ſtomach, and was ſomewhat griped, with noiſe in my bowels. In the evening, and during the night, I paſſed a vaſt deal of wind downwards.

On the ſecond day I was leſs troubled with wind, and on the third and following days hardly at all. My appetite was not perfectly ſatisfied, but my ſpirits were ſomewhat raiſed on the firſt day, and afterwards continued better than when I lived on ſugar.

On the third day of this period I began to have deſires, which were conſiderable in the night.

On the fifth day, Venus ſemel. Having every day, during this period, paid particular attention to the weight of the body, I obſerved that the principal increaſe of weight was on the three laſt days.

EXPERIMENT

EXPERIMENT. VII†.

Diet of Bread, with only the lean Part of boiled Beef, and Water.

	State of the atmofphere during the period.	Daily allowance of food.	Number and total weight of ftools.	Lofs of weight at the end of the period.
First period, three days, Nov. 8, 9, 10.	Thermometer from 49 to 54. Weather rainy and cloudy the 2 firft days; the laft it was ferene.	Bread, 20 oz. The lean ot boiled beef, 12 oz. Water, 3 lb.	1 ftool, weighing 1 lb. 6 oz.	3 lb.
Second period, three days, November 11, 12, 13.	Thermometer from 45 to 47. Weather fair the 2 firft days, rainy the laft.	Bread, 1 lb. The lean of boiled beef, 1 lb. Water, 3 lb.	1 large thin ftool on the morning of the 11th	1 lb.
Third period, three days, November 14, 15, 16.	Thermometer 43. Weather ferene the two firft days, cloudy the laft.	Bread, 9 oz. The lean of * ftewed beef, 18 oz. Water, 3 lb.	5 thin ftools, weighing about 1 lb.	3 lb.

† From this time I made ufe of the Avoirdupois weight only.

* *Although the beef is faid to have been ftewed during the third period, this circumftance does not feem to me to make any alteration in the Experiment, as Dr. Stark did not ufe the gravy, and his meat was but badly cooked.*

REMARKS.

REMARKS.

My appetite was by no means ſatisfied during either the firſt or ſecond period. I ſcarcely paſſed any wind either way. My ſleep was ſomewhat diſturbed by dreams. I had ſtrong deſires. (Venus bis)

On the firſt day of the laſt period, before I was attacked with the purging, my appetite was hardly ſatisfied with a meal, conſiſting of eight or ten ounces of beef, and about half as much bread—I became hungry a few hours afterwards, had frightful dreams in the night, and awoke ſeveral times with palpitation at the heart.

Having obſerved ſome pieces of the beef paſs through me undigeſted, I imagined that the purging was owing to the beef I had ate, being tough and badly dreſſed; for I had not yet learnt the time that was neceſſary to prepare it properly.

By repeated trials I found, that ſix or ſeven hours of the boiling heat was neceſſary to make the beef tender; that by this time one third of the meat which was put into the inner pan, without any water, was gravy, or a fluid, which congealed on cooling, whilſt two-thirds only remained ſolid. In preparing ſeveral pounds of meat at a time, there was only the loſs of a few drachms in the weight, which, I imagine, was chiefly air, as I obſerved many air-bubbles to ariſe through the gravy. Finding it impoſſible to ſeparate entirely, all the fat from the lean, when raw,

raw, the oil which roſe to the ſurface in preparing the beef, was, when cold, carefully removed.

N. B. Dr. Stark, during the two firſt periods of this Experiment, had the boiled beef from an eating-houſe, but for the laſt three days it was dreſſed at home, in a cloſe veſſel, of which he gives the following deſcription. " The veſſel in which the beef was cooked, and which I employed afterwards in preparing all my food, was a tin pan, of a cylindrical form, about three inches in diameter, and capable of containing about three pints and a half, wine meaſure; this pan had a cloſe cover, and was ſuſpended in another of the ſame ſhape, about two inches deeper and wider; the intermediate ſpace being filled with water, before the veſſel was put on the fire; the inner pan was, by this means, a kind of oven or balneum Mariæ, in which the heat was always equal, and the air excluded."

EXPERIMENT VIII.

Diet, ſtewed Lean of Beef, with the Gravy and Water.

Period, four days, November 17, 18, 19, 20.	State of the atmoſphere during the period.	Daily allowance of food.	Number and total weight of ſtools.	Loſs of weight at the end of the period.
	Thermometer from 39 to 40. Weather, for the 3 firſt days ſerene or cloudy, on the laſt rainy.	Stewed beef, 20 oz. beſides the gravy. Water 3 lb.	1 ſoft ſtool on the 19th weighing 3 oz. 7 dr.	2 lb.

REMARKS.

In two or three hours after a meal of ten or twelve ounces of meat with its gravy, I became hungry, and was particularly ſo every night at bed-time. I never had any wind in my ſtomach, and very ſeldom paſſed any downwards. My ſpirits, at all times very good, were ſomewhat raiſed after each meal; but my ſleep was every night diſturbed by dreams, a circumſtance which was new to me. I commonly awoke very early in the morning, and found myſelf lively and well refreſhed: and although I had not ſlept my uſual time, I was never drowſy of an evening. I had ſometimes weak deſires at the beginning of this period, but none afterwards. My ſtools reſembled in colour, the ruſt of iron.

N. B. I tried at this time to leſſen my uſual allowance of water; an experiment which I had ſometimes made before, but I found that it could not be done without occaſioning great thirſt.

EXPERIMENT IX.

Diet, ſtewed Lean of Beef with the Gravy, Oil of Fat or Suet, and Water.

Period	State of the atmoſphere during the period.	Daily allowance of food.	Number and total weight of ſtools.	Loſs of weight at the end of the period.
Firſt period, three days, November 21, 22, 23.	Thermometer from 43 to 46. Weather variable, the 22d rainy.	Stewed beef, 20 oz. beſides the gravy, Oil of fat, 7 oz. Water 3 lb. 4 oz.	1 looſe ſtool, weighing 10 oz. 7 dr.	1 lb. 2 oz. 8 dr.
Second Period, November 24.	Thermometer 43. Weather cloudy, with rain.	Stewed beef, 20 oz. Oil of fat, 9 oz. Water 3 lb.	2 looſe ſtools, weighing 1 lb.	1 lb. 1 oz.
Third period, Novembe 25.	Thermometer 48. Weather cloudy, with rain.	Stewed beef, 24 oz. Oil of ſuet, 1 oz. Water, 3 lb.	1 thin ſtool, weighing 8 oz. 8 dr.	7 oz.

REMARKS.

REMARKS.

Having already aſcertained the nutritious quality of olive or vegetable oil, joined with bread, I was deſirous of trying if animal oil, when taken with the lean part of meat would have a ſimilar effect. The firſt day I took only four ounces of oil, obtained from the common, or outſide fat. The ſecond day I took ſix ounces, and the third day I took ten ounces of oil procured from ſuet. It did not diſagree with my ſtomach, although it was not intimately mixed with the ſoup*, but floated on the ſurface of it— I, however, had ſome wind in my ſtomach; and, being thirſty, was obliged to encreaſe my uſual quantity of water.—I ſlept longer, and more quietly than formerly, and was more diſpoſed to be drowſy than when I lived on the lean of meat only.

N.B. I found that of beef ſuet, ſeven-eighths was pure oil or tallow, whereas the common, or outſide fat, did not yield above two-thirds of oil, one-third being mucilage or cellular ſubſtance. The mucilage diſſolved readily in water, and formed a jelly with it, but both mucilage and cellular ſubſtance, when ſeparated from the oil, were extremely offenſive to the ſmell and taſte.

* *Dr. Stark's ſoup was a little warm water, added to the gravy of the meat.*

EXPERIMENT X.

Diet of Flour, Oil of Suet, Water and Salt.

Period	State of the atmosphere during the period.	Daily allowance of food.	Discharges by stool and urine.	Encrease of weight at the end of the period.
First period, five days, Nov. 26, 27, 28, 29, 30.	Thermometer from 45 to 48. Weather, much rain on the 26th, the other days serene or cloudy.	Flour, 20 oz. Oil of suet, 6 oz. Water, 4 lb. Salt, 12 dr.	2 soft stools, weighing 9 oz. 12 dr.	7 lb. 15 oz. 13 dr.
Second period, two days, December 1, 2.	Thermometer 45 and 43. Weather serene, or cloudy.	Food as above.	Urine, 5 lb. 13 oz. 2 soft stools, weighing 1 lb. 10 oz.	Loss of weight, 3 lb. 14 oz. 13 dr.
Third period, December 3.	Thermometer 42. Weather cloudy.	No food.	Urine, 3 lb. 15 oz.	Loss of weight 3 lb. 7 oz. 10 dr.
Fourth period, five days, December 4, 5, 6, 7, 8.	Thermometer from 41 to 44. Weather serene, or cloudy, rain on the 6th.	Flour, 20 oz. Oil of Suet, 4 oz. Water, 4 lb. Salt, 12 dr.	Urine, 10 lb, 4 oz. 2 soft stools, weighing 1 lb. 7 oz. 5 dr.	Encrease of weight 4 lb. 11 oz. 6 dr.

REMARKS.

REMARKS.

I began the preceding Experiment with a view of comparing the nutritious and other qualities of flour with thofe of the lean of meat. The quantity of tallow ufed in both Experiments was nearly the fame, the quantity of water was regulated by the thirft, and varied from 3¼ ℔ to 4½ ℔. In this laft Experiment, the tallow and flour were intimately united, being made into a pudding, with twelve and fometimes twenty ounces of water, the allowance of water ufed as drink being leffened in proportion. On this diet my appetite was fufficiently fatisfied, I was eafy in my bowels and flept very quietly. I obferved, however, that the quantity of fat was too great, as a confiderable part of it paffed through me undigefted in the form of granules. Venus femel during the firft period. I remarked alfo a great inequality in the encreafe of the weight of my body. On the firft day the encreafed weight was 1 ℔ 15 oz.8 dr.—on the fecond, 1 ℔ 15 oz. —on the third,2 ℔ 13 oz. 4 dr.—on the fourth, 10 oz. 4 dr.—on the fifth, 10. 13 dr. This great variation may have been partly owing to the retention of the food in the inteftines during the firft days of the period; and on the fecond day I drank more water than ufual, which, probably caufed the great encrea weight on the morning following.

During the fecond period I found the diet begin to difagree with me; I loft my appetite, and was feized with fevere head-achs, with uneafinefs at my ftomach and bowels, and great part of the tallow paffed through my body unaffimilated. I was thirfty, and greatly troubled with wind, both upwards

wards and downwards. I alſo at this time obſerved a conſiderable encreaſe in my urine.

Having been extremely uneaſy during the night of the ſecond of December, and having no appetite for food on the morning of the third, I thought proper, though my appetite returned in the afternoon, to abſtain from food the whole day, and next morning was quite well.

Suſpecting that the bad effects of the preceding diet were owing to the quantity, and not the quality of the tallow, I diminiſhed the quantity during the laſt period, and had then the ſatisfaction to find the diet agree with me perfectly well. My bowels were quite eaſy, and I was not troubled with wind, with thirſt, or with head-ach, and no part of the tallow remained undigeſted.

The weight of my body was encreaſed on the firſt day, 2 ℔ 14 oz. 8 dr.—on the ſecond, 1 ℔ 11 oz.—on the third, having had a large ſtool, there was a loſs of weight 5 oz. 9 dr.—on the fourth, again an encreaſe of 4 oz. 10 dr.—on the 5th, of 3 oz. 7 dr.

I ſhould, poſſibly, have continued longer on this diet, which I found both nouriſhing and agreeable, but wiſhing to aſcertain, as exactly as poſſible, the effect of the oil or tallow, I began the following

EXPERIMENT

EXPERIMENT XI.

Diet of Flour, Water and Salt.

Period, December, 9, 10, 11, 12, 13.	State of the atmosphere during the period.	Daily allowance of food	Discharges by urine and stool.	Loss of weight at the end of the period.
	Thermometer from 42 to 48. Weather various, rain on the 10th and 11th, frost on the 13th.	Flour, 24 oz. Water, 4 lb. Salt, 12 dr.	Urine, 17 lb. 2 soft stools, weighing 1 lb. 1 oz. 13 dr.	5 lb. 6 oz. 5 dr.

REMARKS.

On the first day of this Experiment my appetite was pretty well satisfied, but afterwards, particularly towards the end of it, I found that in two or three hours after a meal, consisting of one half my pudding, I became hungry, and I was extremely so every night at bed-time.

On the former diet, with oil or suet, four pints of water were hardly sufficient to quench my thirst; and, commonly, at bed-time, I was obliged to sip a little more. On the present diet I was never thirsty, and am persuaded that I might at this time, without inconvenience, have diminished my common allowance of water, but I continued it nearly the same for an obvious reason, viz. that I might judge with more accuracy of the effect of the suet or oil joined with the other parts of my food.

When

When the pudding was made with ſuet, I found the one half of it rather too much for one meal; whereas when it was made without ſuet, I ſometimes thought that I could eaſily have ate the whole at one time. Whilſt I lived on a pudding made with oil or ſuet, I felt no inconvenience from retaining my water the whole night, but on the preſent diet, I found it very difficult and even painful to do ſo; and ſeveral times whilſt I was engaged in the morning, in obſerving the nocturnal perſpiration, a little urine run off involuntarily. My fœces were, during this experiment, of an orange colour, during the former of a buff colour, and were of a ſtill lighter colour when the proportion of fat was greater.

I loſt in weight, on the firſt day, 5 oz. 6 dr.—on the ſecond, 10 oz. 3 dr.—on the third, 1 ℔ 3 dr.—on the fourth, 1 ℔ 10 oz. 7 dr.—on the fifth, 1 ℔ 12 oz. 2 dr. thus when the body was not properly nouriſhed, the loſs of weight was greateſt on the laſt days of the Experiment, but when the body was more than ſupported, the encreaſe of weight was greateſt on the firſt days of ſuch a regimen. I alſo remarked, that the encreaſe of urine, was nearly, though not exactly, in proportion to the decreaſe of the weight of the body.

A REPETITION

A Repetition of EXPERIMENT X.

Diet of Flour, Beef Suet, Water and Salt.

	State of the atmoſphere	Allowance of food.	Diſcharges.	Gained in weight.
December 14.	Thermometer 45. Weather cloudy, with rain.	Flour, 24 oz. Suet, 4 oz. Water, 4 lb. Salt, 12 dr.	Urine, 2 lb. 12 oz.	9 oz. 15 dr.

REMARKS.

To aſcertain more fully the effect of ſuet in my pudding, I again repeated it for one day, and obſerved, as formerly, that my appetite was ſatisfied with half the quantity, and that I was not hungry until five hours after my uſual meal. I was a little thirſty after dinner, and my urine was one pint two ounces leſs inquantity than on the preceding day.

EXPERIMENT XII.

Diet of Flour, fresh Butter, Water and Salt.

	State of the atmosphere.	Allowance of food.	Discharges.	Loss of weight.
December 15.	Thermometer 45. Weather rainy and cloudy.	Flour, 24 oz. Butter, 4 oz. Water, 4 lb. Salt, 12 dr.	Urine, 2 lb. 7 oz. 2 liquid stools, weighing 1 lb.	1 lb. 3 oz. 10 dr.

REMARKS.

Finding that the result of the Experiments with suet, or animal oil, corresponded very much with those I formerly made with oil of olives, a vegetable expressed oil, I was desirous of extending my enquiry to other oily substances. I began with fresh butter, which I imagined might safely be taken in the same quantity as suet or oil of beef, but soon after dinner, which was this day my second and last meal, I became uneasy at my stomach, brought up some wind and had pain in my bowels, and soon afterwards had two thin stools, accompanied with considerable heat in the fundament, straining, and even with sweating and trembling. I was extremely ill all the evening, and continued very uneasy in my bowels, and with a pain in my fundament during the whole of the night.

EXPERIMENT

EXPERIMENT XIII.

Diet, Yolks of Eggs, Suet, Figs and Water.

	State of the atmosphere during the period.	Allowance of food.	Discharges.	Loss of weight
December 16.	Thermometer 49. Weather fair and serene.	Yolks of eggs, Suet, of each 4 oz. Figs, 1 lb. Water 4 lb.	Urine, 3 lb. 14 oz. 1 liquid stool, weighing 4 oz. 6 dr.	13 oz. 10 dr.

REMARKS.

Disappointed in the effect of butter, and not having provided any other food, I was this day somewhat irregular; wishing to know the precise effect of flour, and to have some means of judging of the share which it had in the preceding nourishing diet, I had, for some days, been trying to unite or combine fat and water, by means of some mucilaginous substance, imagining that if they could be retained in the body, they would, perhaps, supply a sufficient nourishment without the flour. Gum Tragacanth, which is the strongest vegetable mucilage, a jelly of calves feet, whites of eggs, and the yolks of eggs, were tried, in various proportions, the last answered the best of any, although it did not form a complete union between the suet and water. I breakfasted on the quantity mentioned in the table, with two pints of warm water, imagining that the ſtomach and bowels would poſſible complete the union.

After breakfaſt I became ſomewhat uneaſy at my ſtomach, and in two hours had a liquid ſtool, reſembling exactly the food I had taken, and which contained ſome of the clear melted fat, not united with the water or egg. I had no pain in my bowels, or ſtraining with this ſtool, as with thoſe occaſioned by the butter, and I was ſoon hungry My urine alſo, was greatly encreaſed after the above liquid meal.

I likewiſe tried, if by coagulating the yolks of eggs, and continuing the heat for ſeveral hours, it was poſſible to unite the tallow more intimately, but in this I was diſappointed, and the meſs was ſo diſagreeable, that, after taſting it, though I was extremely hungry, I could not eat it, and therefore dined on one pound of figs, with two pints of tea, which was a very agreeable meal, and I did not become hungry again till after five hours.

EXPERIMENT XII. VARIED.

Diet of Flour, Butter, or Oil of Butter, Water and Salt.

Period, December 17, 18, 19, 20.	State of the atmoſphere during the period.	Daily allowance of food.	Diſcharges.	Gained in weight at the end of the period.
	Thermometer from 44 to 49. Weather, variable, rain on the 18th.	Flour, 24 oz. Butter, or oil of butter, 4 oz. Water, 4 lb.	Urine, 11 lb. 2 oz. 2 thin ſtools, on the 19th, weighing 1 lb. 6 oz. 9 dr.	1 lb. 7 oz. 3 dr.

REMARKS.

Suſpecting, that the butter not having been intimately combined with the flour and water, in the firſt Experiment, was the reaſon of its diſagreeing with my ſtomach and bowels; and being alſo perſuaded, that though in this way, it was found to diſagree with the ſtomach, yet the oil of butter, ſeparated from the other parts, and taken by itſelf, might not have the ſame effect; I was deſirous of aſcertaining both theſe facts: and therefore, in the preceding Experiment, I employed freſh butter, and oil of butter alternately, both of them being mixed up with the flour and water into a pudding.

On the 17th, in the morning, I was quite well, and had a good appetite for breakfaſt, but I had no appetite for dinner, and

and ate my pudding, made with butter, with reluctance. After dinner I was drowzy, thirſty, and obliged to drink half a pint more than my allowance. I had conſiderable uneaſineſs in my bowels, with ſome wind downwards, but no ſtool.

On the 18th, when I uſed the oil of butter, I had a very good appetite for dinner, and no thirſt, or uneaſineſs in my bowels after it, although I paſſed a good deal of wind.

On the 19th, when I again employed butter, I was thirſty, uneaſy in my bowels, and had two looſe ſtools, with ſtrainings and pain in my fundament.

On the 20th, when I made uſe of oil of butter, my appetite was very good, and I had very little thirſt, or uneaſineſs in my bowels, but ſtill I was not quite ſo eaſy as I had been when I uſed the ſame quantity of the oil of ſuet.

EXPERIMENT XIV.

Diet of Flour, Oil of Marrow, Water and Salt.

	State of the atmosphere during the period.	Daily allowance of food	Discharges.	Encreafe of weight.
First Period, December, 21, 22.	Thermometer o Weather o	Flour, 24 oz. Oil of Marrow, 4 oz. Water, 4 lb. Salt, 12 dr.	Urine, 4 lb. 6 oz.	1 lb. 4 oz. 2 dr.
Second period, December 23, 24, 25.	Thermometer o Weather o	Flour, 24 oz. Oil of Marrow, 6 oz. Water, 4 lb. Salt, 12 dr.	Urine, 7 lb. 12 oz. 2 foft ftools, weighing 1 lb. 2 oz. 1 dr.	1 lb. 4 oz. 13 dr.

REMARKS.

Marrow, by gentle heat and preffure, yields about $\frac{11}{12}$ of a pure oil, much pleafanter, both to the tafte and fmell, than the oil obtained from fat or fuet. This oil was combined, as ufual, with flour and water, into a pudding; with which, though my appetite was fufficiently fatisfied, yet I was hungry for

for each meal. I was not in the leaſt thirſty, was eaſy in my bowels, brought up no wind, and paſſed none downwards. I found myſelf remarkably well on this regimen, and thought my ſpirits raiſed by it; though this might be only opinion, as it is difficult on ſuch ſubjects to diſtinguiſh between fancy and reality. I ſometimes had deſires. Venus ſemel, during the firſt period,

Finding the oil of marrow ſo mild in the bowels, and at the ſame time ſo agreeable a food, I encreaſed the quantity, to judge ſtill further of its effects, and particularly to determine whether the degree of nouriſhment, or rather of encreaſe in the weight of the body, was in proportion to the quantity of nouriſhment taken.

I had a ſtool on the 25th, and another on the morning of the 26th, but in neither could I perceive any granules, as was the caſe when I uſed the ſame quantity of fat or ſuet.

I continued perfectly eaſy until the 26th, when I felt myſelf ſomewhat dull before dinner, brought up ſome wind, and had little or no appetite. In the evening I was very drowſy and thirſty, and obliged to drink half a pint more than my common allowance of water, but on this, and even on the preceding day, the angles of ſeveral of the gums were purple, and a little ſwelled. Venus ſemel.

EXPERIMENT

EXPERIMENT XV.

	State of the weather.	Diet irregular.	Difcharges.	Weight of the body in the morning.
December 26.	o	$\frac{1}{3}$ of a pudding made with 6 oz. of fuet. Water, or tea, 2 lb. 6 oz. Black currants, 8 oz.	Urine, 2 lb. 10 oz. 1 ftool, weighing 9 oz.	10 ft. 13 lb. 9 oz. 6 dr.

REMARKS.

As the oil of marrow feemed to be lefs nutritious than that of fuet, I purpofed, in order more exactly to afcertain the fact, to have again taken the fuet for two or three days, particularly as I wifhed to clear up a doubt which I ftill entertained, viz. whether the fame food, or food of the fame nutritious power, taken when the body is in a low ftate, may not raife it fafter than if taken when the body is in better condition. When I firft began to ufe the fuet my body was extremely low, which was not the cafe when I began to ufe marrow, and therefore, to this circumftance, poffibly, may be afcribed, the apparent difference of their nutritious powers. I was, however, prevented making this Experiment, by having no appetite in the morning; and, though I forced myfelf to eat part of a fuet pudding for breakfaft, I could take no more food during the day, and fuffered much uneafinefs

 from

from wind in my bowels. In the evening I was eafier, and ate half a pint of black currants. I was determined alfo, by the appearance of my gums and fkin, to change, for fome little time, my plan of living.

Although upon my pudding diet, I had in general pretty good fpirits, yet I fancied that I was not fo lively as ufual, nor fo active and vigorous, either in body or mind.

N. B. *As Dr. Stark made no abftract of his journal after the 26th of December, the Editor has endeavoured to fupply this lofs from the original journal in his poffeffion.*

EXPERIMENT XVI.

Diet. Bread, with roasted Fowl, Infusion of Tea and Sugar.

Day of the Month,	*State of the Weather	Allowance of food	Discharges by stool and urine.	Weight of my body,
December 27.	Serene, Rainy, Cloudy.	Bread, 2 lb. Roasted fowl, 8 oz. with a little salt, Infusion of tea, sweetened with sugar, 3 lb. 9 oz.	Urine plentiful. 1 large stool.	10 st. 9 lb. 9 oz. 14 dr.
28.	Rainy, Cloudy, Variable.	Bread, 2 lb. Roasted fowl, 12 oz. 3 dr. Tea, 3 lb. 9 oz.	Urine, 3 lb. 3 oz.	10 st. 9 lb. 13 oz. 10 dr.

REMARKS.

Dec. 27. I slept quietly, and awoke this morning, at my usual time, hungry and perfectly easy. Immediately after getting up I had a buff-coloured stool. Was not my indisposition of yesterday occasioned, by my having rather imprudently encreased, and persevered too long in the use of the oil of marrow? which, when taken in a moderate quantity, seems, of all fats, the mildest in the bowels.

* Dr. Stark, from this time, seems to have paid no attention to the thermometer, though he has noticed the weather with particular accuracy.

 This

This morning I obſerve the gums of the double teeth, on each ſide of the upper jaw, conſiderably ſwelled, of a purple colour, and, ſome of them, almoſt black at the corners; they are, likewiſe, hot and painful; thoſe of the left ſide bled on my biting a bit of bread.— The gums of the lower jaw appear to be quite ſound. Moſt of the gooſe-ſkin eminences on my legs and thighs are of a deep red, ſome of them purple; and the diſcolouration, which extends even beyond the eminences, is ſomewhat browniſh at the edges. Under my left breaſt there is alſo a true petechial ſpot, having the ſame appearance, as formerly during my ſevere illneſs.

It is worthy of being remarked; that after I had lived for ſome time, on animal food entirely, although I was reduced lower in weight, in ſtrength, and in ſpirits, than at preſent; yet there were no ſuch appearances. Is it not probable, then, that animal oils, though they nouriſh and encreaſe the weight of the body, are not of themſelves ſufficient, to prevent a morbid alteration from taking place in the blood and fluids? Whilſt, on the other hand, the lean of meat, though leſs nutritious, is of more efficacy in preſerving the fluids of the body in a ſound ſtate? Notwithſtanding, however, what I have obſerved of my gums, and the eminences on my legs; my countenance, and ſkin in general, has the appearance of health.

My food, this day, I found quite ſufficient to ſatisfy my appetite. I had a little wind both upwards and downwards.

28th. When I awoke this morning, I perceived a diſagreeable, ſweetiſh taſte in my mouth, and my gums had an offenſive ſmell; in other reſpects I was much as yeſterday. Towards evening my gums were conſiderably eaſier and better, but I was attacked with a ſevere cholic, which continued moſt of the night.

EXPERIMENT

EXPERIMENT XVII.

Diet. Bread, ſtewed Lean of Beef, with the Gravy, Infuſion of Tea, with Sugar.

Day of the month.	State of the weather.	Allowance of food.	Diſcharges by urine and ſtool.	*Weight of the body.
December 29.	Variable, Serene, Ditto.	Stewed beef, &c. 1 lb. Bread, 2 lb. Tea, 3 lb. 9 oz.	Urine, 3 lb. 7 oz.	10 ſt. 10 lb. 7 oz. 6 dr.
30	Variable, Snow, Rain,	Food, ditto.	Urine, 2 lb. 13 oz. Stool, 7 oz. 4 dr.	10 ſt. 12 lb. 9 oz. 13 dr.
31	Variable, Serene, Ditto.	Food, ditto.	Urine, 4 lb.	10 ſt. 12 lb. 9 oz. 14 dr.
Jan. 1.	Cloudy, Variable, Cloudy.	Food, ditto.	Urine, 3 lb. 15 oz. Stool, 8 oz. 3 dr.	10 ſt. 13 lb. 4 oz. 4 dr.

REMARKS.

* The weight was taken immediately after breakfaſt.

REMARKS.

Dec. 29. I paſſed an unpleaſant night, having been either kept awake by the pain of the cholic, or having my ſleep diſturbed by diſagreeable dreams. In the morning I had not ſo much of the ſweetiſh taſte in my mouth, my gums were paler, leſs ſwelled, and not ſo offenſive to the ſmell as on the preceding day; the ſpots on my ſkin were alſo paler.

My appetite was not ſufficiently ſatisfied with four ounces of meat for breakfaſt, but I found that eight ounces at dinner, and four at ſupper, were rather too much for me. I had no uneaſineſs in my bowels, and paſſed but little wind either way. At bed-time I was thirſty, and drank a few ounces of water.

30. I ſlept quietly until an hour or two before day-light, when I had a little diſturbance in my bowels, but without pain, My gums now ſcarcely occaſion any offenſive ſmell or taſte. At dinner, beſides my uſual allowance of beef, &c. I ate ſome rice-pudding, with melted butter, and drank two glaſſes of wine. After dinner I had a pain at the pit of my ſtomach, but which went off upon my bringing up a little wind. Before ſupper I had a head-ach, but this went off; alſo, after a firm ſtool, of a dark, earthy colour, attended with violent ſtraining; a little before going to bed I was troubled with wind, and a good deal griped.

31. My appetite is ſatisfied with my preſent allowance of food, which I find would not be the caſe were I at all to leſſen it. I commonly

monly eat my beef cold, being more agreeable to me than when hot My bowels are quite easy, I passed a little wind downwards, but none upwards. My gums, though not livid as at first, are still red, a good deal puffed up, are apt to bleed on being pressed with the finger, and are so painful, that it is troublesome to me to eat even the crumb of bread. This evening I perceive that the spots on my skin are paler than they were in the morning. Although the quantity of my drink was the same as formerly, my urine is to-day considerably encreased. I observe that the urine, collected during the night, is much paler than what I make in the day.

Jan. 1. Although I sleep quietly every night, yet my gums are still puffed up, livid and uneasy; and in the left side, there is a small smarting sore, in a place from which a tooth was extracted some years ago. To-day I have been hungry for each meal, and was considerably so after supper. I am quite easy and in good spirits, with little or no wind either way; before dinner I had a firm stool of a uniform dark brownish colour. Although, at present, I take more food than what is absolutely necessary for the support of the body, I remain perfectly well, whereas I have several times suffered considerable inconvenience from committing any excess in the quantity of oils. Is it not evident, then, that an excess in the use of oils, is more hurtful to the body, than an excess in any other article of food? and that, of course, we ought to be particularly careful in regulating the quantity and quality of the oils we employ in diet.

EXPERIMENT XVIII.

Diet. Bread, The Fat of ſtewed Beef, with the Jelly, Water and Salt.

Day of the month.	State of the Weather.	Allowance of food.	Diſcharges by urine and ſtool.	Weight of the body.
Jan. 2.	Cloudy, Ditto, Rain.	Fat of beef, 4 oz. Bread, 2 lb. Water, 3 lb. 9 oz.	Urine, 4 lb. 6 oz.	10 ſt. 12 lb. 10 oz. 14 dr.
3.	Cloudy, Ditto, Ditto.	Fat of beef, 4 oz. Bread and water as yeſterday.	Urine, 3 lb. 10 oz. Stool, 10 oz. 6 dr.	10 ſt. 13 lb. 6 oz. 2 dr.
4	Cloudy. Ditto, Snow.	Food, ditto.	Urine, 3 lb. 10 oz. Stools, 11 oz. 6 dr.	10 ſt. 13 lb. 6 oz. 2 dr.
5	Serene, Froſt, Ditto.	Food, ditto.	Urine, 2 lb. 12 oz.	10 ſt. 13 lb. 6 oz. 2 dr,

REMARKS.

REMARKS.

Jan. 2. With a view to compare the effects of the fat of beef, (which may be ſomewhat different from the oil already tried) with thoſe of the lean, I began, this day, a courſe of Experiments with the former. The fat which I uſed, was ſtewed with the lean of the beef, ſeparated from it when cold, and ate, with as much of the jelly as ſeemed to belong to it; the pure oil, which had concreted on the ſurface, was entirely removed. But, as the lean of ſtewed beef had ſome fat mixed with it, in like manner the fat has ſome lean, which renders the Experiment not ſo complete as one I purpoſe making with boiled buttock of beef, of which I ſhall uſe the two parts accurately ſeparated from each other; but, previous to this, I intend to try the effect of greens and freſh fruit.

Is it not poſſible, that a ſmall quantity of fat may afford the ſame nouriſhment as a larger quantity of lean, and may be ſo prepared as to be more eaſily aſſimilated by weak digeſtive organs? Perhaps marrow, mixed up with panada, might prove a proper and uſeful food for convaleſcents.

This morning my gums were quite well, and the gooſe-ſkin eminences on my legs were only of a pale red. I was hungry for breakfaſt, and made a very agreeable one, upon two ounces of fat*, with bread and tea; and, finding myſelf hungry ſooner than I expected after breakfaſt, I took, for dinner, three ounces of fat, which was alſo a very agreeable meal. In the evening I ſupped on bread and tea. I was in excellent ſpirits, much inclined to venery, to which I did not feel the ſmalleſt propenſity, whilſt living on the lean of meat. I paſſed little wind either way.

3. (Venus bis.) My ſleep was ſomewhat diſturbed in the night by diſagreeable dreams; my appetite, to-day, has been very well ſatisfied until evening, when I felt myſelf hungry,

4. I was reſtleſs laſt night, and had a frightful dream. A few hours after breakfaſt I had a firm, dark brown ſtool. The affection of my gums ſtill continues, although in a ſlighter degree than when I began to eat meat.

5. I ſlept well in the night, excepting that I was diſturbed by a dream, attended with an emiſſion, a circumſtance that has not happened to me above three or four times in my life. I was rather dull after breakfaſt, probably owing to my having ate too much fat.

A Repetition of EXPERIMENT VII.

Diet. Bread, the Lean of boiled Beef, Infusion of Tea with Sugar, (no Salt.)*

Day of the month.	Sate of the weather.	Allowance of food.	Difcharges by urine and ftool.	Weight of the body.
Jan. 17.	Serene, with froft.	Bread, 2 lb. Lean of beef, 1 lb. Infufion of tea, 3 lb. 9 oz.	1 ftool, weighing 4 oz. 4 dr.	11 ft. 3 lb. 13 oz. 4 dr.
18.	Cloudy, Snow, Rain.	Food, ditto.	Urine, 4 lb. Stool, 1 lb. 4 oz. 12 dr.	11 ft. 1 lb. 8 oz. 6 dr.
19.	Rain, Serene, Cloudy.	Beef, 12 oz. Bread, as above.	Urine, 2 lb. 14 oz.	11 ft. 11 lb. 12 oz. 10 dr.
20.	Cloudy, Ditto, Ditto.	Beef, 14 oz. Bread, as above.	Urine. 3 lb.	11 ft. 3 lb. 1 oz. 8 dr.

 REMARKS.

* Sir John Pringle' having a fufpicion that the large quantity of falt which I formerly ufed, might occafion the affection of my gums; I omitted it entirely in the prefent Experiment.

REMARKS.

Having, ſince the 5th, been engaged in a canvas for St. George's Hoſpital, I have been very irregular in reſpect to diet, living rather more freely, and drinking more wine than uſual. I have alſo walked a good deal; but, upon the whole, have been extremely well in health, excepting one night, when, from a deſire of preventing hunger the next morning, I ate too much fat for ſupper. I have had commonly one ſoft ſtool a-day. My gums are quite firm and well, and the gooſe-ſkin eminences are of the natural colour of the ſkin. The weather has been, in general, ſevere, with froſt and ſnow, until within theſe few days, when it has been a little milder.

17. My appetite, this day, has been rather more than ſatisfied. I paſſed a good deal of wind downwards, and, before ſupper, had a ſmall firm ſtool.

18. Early this morning I felt conſiderable uneaſineſs in my bowels, had a ſtool, the latter part of it thin. Had no appetite for ſupper, which I would rather have omitted; very dull all day; took [no exerciſe till the evening, had then a good deal of pain in my belly, and was greatly troubled with wind.

19. I was ſtill low-ſpirited and dull, but this was, poſſibly, in part, owing, to my having but ſmall hopes of ſucceſs at St. George's

George's Hofpital. I had fome difturbance in my bowels in the morning, and again in the evening, attended with pain. My appetite was rather more than fatisfied, and I had fome flight defires.

20. I went to bed foon after fupper, dreamt in the night, though not frightfully; paffed a good deal of wind before I got up; ufed a good deal of exercife in the morning, and breakfafted three hours later than ufual. I find my appetite more than fatisfied, and I am now quite eafy in my bowels.

EXPERIMENT XIX.

Diet. Bread, Fat of boiled Beef, Infusion of Tea with Sugar.

Day of the month.	State of the weather.	Allowance of food.	Discharges by urine and stool.	Weight of the body.
Jan. 21.	Cloudy, Variable, Serene.	Fat of boiled beef, 3 oz. Bread, 2 lb. Tea, 3 lb. 9 oz.	Urine, 2 lb. 9 oz. Stool, 9 oz. 12 dr.	11 st. 4 lb. 9 oz. 12 dr.
22.	Cloudy, Serene, Rainy.	Fat, 2 oz. Bread, &c. ditto.	Urine, 3 lb.	11 st. 3 lb. 3 oz. 13 dr.
23.	Cloudy, Serene, Ditto.	Food, ditto.	Urine, 1 lb. 13 oz.	11 st. 4 lb. 10 oz. 10 dr.
24.	Serene, Ditto, Ditto.	Fat, 2½ oz. Bread, &c. ditto.	Urine, 2 lb. Stool, 3 oz. 8 dr.	11 st. 6 lb. 2 oz. 3 dr.

REMARKS.

REMARKS.

Jan. 21. My morning urine was high-coloured, and became turbid on ſtanding. This morning I walked a good way before breakfaſt, and breakfaſted an hour earlier than yeſterday; my bowels were ſomewhat uneaſy, and I paſſed a great deal of wind downwards. The fat I uſe has been kept ſome days ſince it was boiled, but ſeems quite ſweet; I found two ounces rather too much for breakfaſt, and I was under the neceſſity of eating half my loaf with it.

Before dinner I had a dark coloured ſtool, of a moderate conſiſtence; at dinner I could not finiſh an ounce of fat. I had a ſlight pain in my bowels after breakfaſt; and, although I had no wind, was a little griped before I went to bed. In the night, after ſome ſevere griping pains, I had a purging, dark-coloured, ſlimy ſtool, which I apprehend to have been occaſioned from the fat (as it had been kept ſome days,) being ſomewhat rancid, although this was not perceptible either to the ſmell or taſte. Poſſibly, the ſudden change of diet was partly the cauſe of it.

22. At breakfaſt I could eat only one ounce of fat, and that with reluctance, as the fat was ſoft and greaſy, which is more diſagreeable than when firm and dry—I ate another ounce at dinner, and although, from being harder, I ate it with leſs reluctance, yet I found it fully enough for my ſtomach. As my appetite

appetite was fully ſatisfied, I ate no ſupper. I was eaſy in my bowels and well in every reſpect. (Had deſires.)

23. (Venus ſemel.) I ate an ounce of hard fat for breakfaſt, without reluctance, it was quite freſh, having been boiled only two days before—I eaſily ate the ſame quantity at dinner—I was perfectly eaſy in my bowels, had no wind upwards, and little downwards—Was ſomewhat hungry at bed-time.

24. The freſh and friable fat became at laſt to be almoſt as agreeable to me as butter. My appetite was not more than ſatisfied, and whilſt uſing this diet I felt myſelf lighter, more alert and eaſier, than when living on the lean of meat only. I had ſcarcely any wind in my ſtomach—At twelve I had a light yellow ſtool of a moderate conſiſtence.

EXPERIMENT XX.

Diet. Bread, the Lean of roaſted Veal, and Water.

Day of the month.	State of the weather.	Allowance of food.	Diſcharges by urine and ſtool.	Weight of the body.
Jan. 25.	Cloudy, Ditto, Rain.	Lean of roaſted veal, 12 oz. Bread, 2 lb. Water, 3 lb. 9 oz.	Urine, 2 lb. 14 oz.	11 ſt. 5 lb. 12 oz. 4 dr.
26.	Rain, Cloudy, Serene.	Food, ditto.	Urine, 4 lb. 5 oz.	11 ſt. 4 lb. 7 oz. 10 dr.
27.	Cloudy. Ditto, Ditto.	Breakfaſt, roaſted veal, 3 oz. Dinner and ſupper irregular.	Urine, 3 lb. 14 oz. Stool, 5 oz. 8 dr.	11 ſt. 3 lb. 15 oz. 14 dr.

REMARKS.

Jan. 25. I was very hungry for dinner, and immediately before it had a ſtool, partly coſtive and partly looſe. May not a ſudden change of diet have an effect in opening the body, even though the change made, be from a kind of food naturally more opening, to one that is leſs ſo? At bed-time I felt ſome commotion in my bowels, and paſſed ſome wind downwards.

26. I have dreamt for ſome nights paſt. To-day I have brought off wind, and ſeveral times ſome ſtuff from my ſtomach; I have likewiſe been drowſy, eſpecially after dinner; I am eaſy in other reſpects, and my appetite not more than ſatisfied.

37. I had deſires in the night, but which went off upon emptying my bladder. I dined and ſupped abroad, and ate heartily of a variety of different things, but, though I did not overload my ſtomach, I was not ſo eaſy as uſual for ſome time after I went to bed.

EXPERIMENT

EXPERIMENT XXI.

Diet. Bread, Fat of Bacon Ham, Infusion of Tea, with Sugar.

Day of the month.	State of the weather.	Allowance of food.	Discharges by urine and stool.	Weight of the body.
January 28.	Cloudy, Ditto, Ditto.	Fat of boiled ham, 1½ oz. Bread, 2 lb. Tea, 3 lb. 9 oz.	Urine, 2 lb. 12 oz. Stool, 10 oz.	11 st. 6 lb. 5 oz. 4 dr.
29.	Rain, Cloudy, Ditto.	Fat of ham, 2 oz. Bread, &c. as above.	Urine, 2 lb. 4 oz. Stool, 1 lb. 6 oz. 8 dr.	11 st. 6 lb. 9 oz. 8 dr.
30.	Cloudy, Mixed, Cloudy.	Fat of ham, 3 oz. Bread, &c. as above.	Urine, 2 lb. 13 oz.	11 st. 5 lb. 12 oz. 4 dr.
31.	Rainy, Ditto. Cloudy,	Fat of ham, 2½ oz. Bread, &c. as above.	Urine, 2 lb. 15 oz.	11 st. 5 lb. 13 oz. 4 dr.

REMARKS.

Jan. 28. The fat of bacon ham is much more greaſy than that of beef. I ate as much of the fat, quite pure, at breakfaſt and dinner, as was agreeable to my ſtomach; and, though at ſupper I was rather hungry, I took none, being apprehenſive of its purging me. Some little time before dinner I had a ſtool of a moderate conſiſtence, covered with a white frothy liquid. At bed-time I was very hungry.

29. Having obſerved that my urine ran off pretty copiouſly, both after breakfaſt and after dinner, and being ſomewhat hungry towards nine o'clock, I intended to have ate half an ounce more of fat at ſupper, but was prevented by a ſudden commotion in my bowels, which was followed by a looſe, frothy ſtool, though without any conſiderable uneaſineſs or griping, as was the caſe after eating butter; I was the whole day uncommonly dull and low-ſpirited, and in the evening, before going to bed, made two pints of water.

30. May not the purgative quality of the fat be imputed, in a great meaſure, to its rancidity?

31. I found no inconvenience from the encreaſed quantity of fat which I ate this morning at breakfaſt. At bed-time I was ſomewhat hungry, and had a looſe ſtool, without any conſiderable uneaſineſs, although I was griped after it; there were ſome whitiſh particles mixed with the ſtool, and along with it alſo ſome ropy mucus. Upon my preſent diet I was never thirſty, had hardly any wind, and no deſires.

EXPERIMENT XXII.

Diet. Bread, Lean of Bacon Ham, Infuſion of Tea, with Sugar.

Day of the month.	State of the weather.	Allowance of food.	Diſcharges by urine and ſtool.	Weight of the body.
Feb. 1.	Mixed, Serene, Ditto.	Lean of ham, 10 oz. Bread, 2 lb. Tea, 3 lb. 9 oz.	Urine, 2 lb. 4 oz.	11 ſt. 5 lb. 13 oz. 2 dr.
2.	Serene, Cloudy. Ditto.	Food, ditto.	Urine. 3 lb. 9 oz. Stool, 10 oz. 8 dr. Ditto, 2 oz. 8 dr. Ditto, 9 dr. Ditto, 3 dr.	11 ſt. 3 lb. 14 dr.
3.	Cloudy, Ditto, Ditto.	Lean of ham, 9 oz. 4 dr. Bread, &c. as above.	Urine, 2 lb. 9 oz. Stool, 2 oz. 6 dr.	11 ſt. 5 lb. 4 dr.

REMARKS.

Feb. 1. I was extremely thirſty after dinner, but was rather hungry than thirſty after ſupper. In the evening I paſſed ſome wind downwards.

2. I awoke

2. I awoke early in the morning with pain in my bowels, paſſed ſome wind downwards, was obliged to get up to the chair, and had a looſe ſtool, of a yellow colour, and unequal conſiſtence. I was very hungry for breakfaſt, and immediately after it was griped, and had a ſecond looſe ſtool, which was ſlimy and accompanied with violent ſtraining; the purging continued all day, attended with great ſoreneſs, pain in my bowels, and violent ſtraining; the ſtools were chiefly ſlime or jelly, with ſome feculent matter and blood; notwithſtanding this indiſpoſition I was hungry at bed-time.

3. I was pretty eaſy during the night, and continued ſo till after dinner, when the uneaſineſs in my bowels returned, and I had a looſe ſlimy ſtool, and paſſed a good deal of wind. My urine was high-coloured. Was the purging owing to the ſalts in the ham, to the firmneſs of its texture, or to its being ſomewhat rancid?

EXPERIMENT XXIII.

Diet. Bread, or Flour, with Honey, and Infusion of Tea or of Rosemary.

Day of the month.	State of the weather.	Allowance of food.	Discharges by urine and stool.	Weight of the body.
Feb. 4.	Cloudy, Ditto, Mixed.	Breakfast, Honey, 3 oz. Bread and tea, Dinner irregular.	Urine, 2 lb. 10 oz. 1 soft stool, not ½ an oz.	11 st. 5 lb. 7 oz. 4 dr.
5.	Serene, Mixed, Rainy.	Honey, 8 oz. Flour, 1 lb. 8 oz. Water, 8 oz. (made into a pudding.) Tea, without sugar, 3 lb. 8 oz.	Urine, 1 lb. 15 oz, Stool, 7 oz.	11 st. 6 lb. 2 oz. 2 dr.
6.	Serene, Ditto, Ditto.	Honey, 8 oz. Bread, 2 lb. Weak infusion of rosemary, 3 lb. 8 oz.	Urine, 4 lb. 10 oz. Stool, 10 dr.	11 st. 3 lb. 14 oz. 3 dr.
7.	Rainy, Snow, Mixed, very cold.	Honey, 12 oz. Flour and water, as above, (made into a pudding.) Infusion of rosemary, 3 lb. 8 oz.	Urine. 1 lb. 15 oz. Stool, 9 oz. Ditto, 10 oz. 8 dr.	11 st. 5 lb. 9 dr.
8.	Serene, Ditto, Ditto, } Frost.	Honey, 6 oz. Bread, 2 lb. Infusion, 3 lb. 8 oz.	Urine, 4 lb. 11 oz. Stool, 6 oz. 4 dr. Ditto, 2 oz. 13 dr.	11 st. 1 lb 13 cz. 6 dr.
9.	Serene, Ditto, Cloudy. } Frost.	No Honey, Bread, 2 lb. Infusion, 3 lb. 8 oz.	Urine, 1 lb. 3 oz.	11 st. 11 lb. 10 oz. 8 dr.

EXPERIMENT

EXPERIMENT XXIII. CONTINUED.

Day of the month.	State of the weather.	Allowance of food.	Difcharges by urine and ftool.	Weight of the body.
Feb. 10.	Rainy. Cloudy, Ditto.	Honey (which had been expofed to a boiling heat) 4 oz. Bread and infufion of rofemary, as formerly.	Urine, 1 lb. 9 oz. Stool, 5 oz. 2 dr.	11 ft. 10 lb. 10 oz. 2 dr.
11.	Cloudy, Ditto, Ditto, } Mild.	Virgin honey, 4 oz. Bread and infufion as formerly.	Urine, 2 lb. 1 oz. Stool, 1 lb. 1 oz.	11 ft. 4 lb. 14 oz. 9 dr.
12.	Cloudy, Ditto, Ditto. } Mild.	Honey (heated in balneo Mariæ) 4 oz. Bread and infufion, as formerly,	Urine, 2 lb. 12 oz.	11 ft. 4 lb. 7 oz. 2 dr.
13.	Mixed, Cloudy, Serene. } Mild	Honey. 4 oz. Flour, 1 lb. 8 oz. Water, 12 oz. (made into a pudding, and ftewed) for feveral hours. Infufion, 3 lb. 4 oz.	Urine, 2 lb. 10 oz.	11 ft. 4 lb. 11 oz. 9 dr.
14.	Serene, Mixed, Serene. } Very mild.	Honey heated, 8 oz. Bread, 2 lb. Infufion, 3 lb. 8 oz.	Urine, 4 lb. 13 oz.	11 ft. 11 lb. 15 oz. 4 dr.
15.	Cloudy, Ditto, Ditto. } Fogg	No Honey, Bread, 2 oz. Infufion of rofemary, 2 lb. 6 oz.	Urine, 2 lb. 6 oz. Stool, 1 lb. 6 oz. Ditto.	11 ft. 11 lb. 8 oz. 10 dr.

REMARKS.

REMARKS.

Feb. 4. I breakfaſted on three ounces of honey with bread; at dinner I was irregular, and drank ſome wine.

5. (Venus ſemel.) I had a ſtool immediately after breakfaſt, of a proper conſiſtence, but which contained ſome pieces of plum and currant-ſkins, which I had ate the preceding day. My honey pudding, which had been ſtewed for ſeveral hours, was ſo firm that I had ſome difficulty in chewing it; a pound of it was rather too much for breakfaſt, and, though I was very hungry at dinner, I found even then a pound more than agreeable. In the afternoon and evening I paſſed ſome wind downwards, at bed-time I was extremely hungry. May we not reaſonably ſuppoſe that food which is difficult to chew is difficult alſo to digeſt? Is not bread of more eaſy ſolution in the ſtomach than pudding, made with the ſame quantity of water. The pudding made with honey, beſides being tough, was, in other reſpects, far leſs pleaſant than bread and honey.

6. This day I varied the Experiment, to try whether the heat, or the intimate combination of the honey with the flour, made any ſenſible alteration in its effects. I made a larger quantity of urine, and which, of courſe, was much paler. Before dinner I had a ſmall ſtool; after ſupper I paſſed ſome wind, and felt ſome ſlight commotion in my bowels; at bed-time I was extremely hungry.

7. In my pudding to-day, only eight ounces of honey were mixed with the paſte, four ounces were added afterwards. Immediately after breakfaſt, I had a ſtool of a common conſiſtence, and before ſupper had a looſe one, but without being griped. This evening I felt rather more commotion in my bowels than on the preceding one. I was very hungry for every meal, and, at going to-bed, extremely ſo.

8. This morning, ſoon after getting up, I was a little griped, and had a looſe ſlimy ſtool of a moderate conſiſtence. During the day I had three more purging ſtools, and was a good deal griped, with conſiderable uneaſineſs in my bowels; I had no appetite for food and was liſtleſs, drowſy, and uneaſy all the evening.

9. I was ſomewhat uneaſy in the night, and this morning early I was obliged to get up to the chair, and had a looſe ſlimy ſtool, about eight ounces in weight. Yeſterday my urine, after ſome time, became turbid, and depoſited a brick-coloured ſediment; to-day it was high-coloured, and became turbid alſo when cold. As I was ſtill ſomewhat uneaſy in my bowels, I thought it prudent to take no honey to-day; I omitted it alſo more readily, imagining that by ſo doing, I ſhould be able to judge more accurately of its effects. I had little wind in my ſtomach during this or the two preceding days, nor had I any acute pain, or griping in my bowels, yet I was dull, and felt a general uneaſineſs. To-day, after taking a walk, I was hungry for dinner, and this morning I perceived, for the firſt time, on the inſide of my cheek, a ſmall, ſmarting, aſh-coloured ulcer, its edges very red and ſwelled, but the gums and ſkin have, as yet, no morbid appearance.

10. (Venus ſemel.) I had ſome uneaſineſs in my bowels in the morning. Being deſirous of aſcertaining the effect of heat on honey, what I uſed to-day was previouſly kept, for three or four hou rs,in balneo Mariæ.

11. After breakfaſt I had a ſtool of the common conſiſtence; iu the evening had ſome commotion in my bowels. The edges of the ſore in my mouth were not ſo much ſwelled as the day before, Was not the retention of urine on the 9th and 10th to be aſcribed, rather to an indiſpoſition occaſioned by uſing too great a quantity of honey, than to the honey itſelf?

12. I was not very hungry either yeſterday or to-day; now and then I was a little griped; in the evening my gums, particularly on the inſide, were hot and ſomewhat ſwelled, a beginning ſcorbutic ſymptom; at bed-time I was again a little griped, and had a ſoft, or rather a looſe ſtool.

13. Having found that heated honey, taken with bread, is not more diuretic than common honey, I again made it into a pudding, to try whether in this way it would not have the ſame diuretic quality as it had in the beginning of theſe Experiments. My urine run off very faſt, and I was extremely hungry at bed-time; I had neither griping, wind, or inclination to ſtool.

14. I was extremely hungry for breakfaſt. I ate a larger quantity of heated honey than I had ever done, to try if it would prove diuretic, by encreaſing the quantity.

Do not the preceding Experiments ſhew, that heated honey, though leſs purgative, is not much more diuretic than virgin-honey? and, as neither bread nor pudding have, of themſelves, any diuretic quality, we are at a loſs to account for the remarkable diuretic effect of honey pudding.

Upon the honey diet I had no deſires, no wind upwards, and little downwards; my ſpirits were, as uſual, pretty good, and my body ſufficiently active.

EXPERIMENT

EXPERIMENT XXIV.

Diet. Bread, with Cheſhire Cheeſe, and Infuſion of Roſemary.

Day of the Month,	State of the Weather	Allowance of food.	Diſcharges by ſtool and urine.	Weight of my body,
February 16.	Cloudy, Ditto, Ditto.	Cheſhire Cheeſe, 4 oz. Bread, not quite 2 lb. Infuſion of roſemary, 2 lb. About 1 lb. of mulled Port.	Urine, Stool, Ditto.	11 ſt. — 5 oz. 12 dr.
17.	Rainy, Cloudy, Ditto.	Bread, 2 lb. Cheeſe, 4 oz. Infuſion of roſemary, 3 lb. Water, 1 lb. 8 oz.	Urine, 2 lb. 4 oz. Stool, Ditto.	11 ſt. — 11 oz. 2 dr.
18.	Mixed Rain & Snow. } Hurricanes.	Bread, with infuſion of roſemary, no cheeſe.	Urine, 1 lb. 10 oz.	11 ſt. 13 lb. 5 oz. 10 dr.

REMARKS.

On the evening of the 14th I was very well when I went to bed, but awoke before day with conſiderable uneaſineſs in my bowels, and had ſeveral looſe ſtools.

In

In the morning of the 15th I was chilly, ſometimes with ſhivering, was liſtleſs and uneaſy, though the uneaſineſs was chiefly in my bowels; I had not the ſmalleſt appetite for food. For breakfaſt I took about two ounces of bread, with a pint of infuſion of roſemary, which, in about an hour after I had taken it, run from me by ſtool without pain. I continued all day extremely uneaſy, ſighing and moaning. Owing to my feebleneſs, I lay moſt of the time in bed, but without being ſenſible of any relief. In the evening, being thirſty, I drank another pint of infuſion of roſemary. In the afternoon, beſides my other complaints, a head-ach came on, which continued all night. During the night I was reſtleſs, very uneaſy in the lower part of my belly, and had five or ſix liquid ſtools, but did not make above a few ſpoonsful of urine.

16. Towards morning the head-ach went off, but I was ſtill uneaſy in my belly, and had no appetite for food. My ſkin retained a natural appearance, and my gums, ſo far from being affected in the manner they had been by ſugar, were univerſally very pale, almoſt white, and not in the leaſt puffed up or painful. Immediately after breakfaſt I had a ſmall watery ſtool; in the forenoon had a good deal of uneaſineſs in my bowels, at times ſome wind upwards. I was quite low and unfit for ſtudy; before dinner I had a ſmall liquid ſtool. In the evening, being ſtill uneaſy in my bowels, and with noiſe in them, I took ſome mulled Port wine, and found myſelf better after it.

Does not an exceſs in ſweets give a ſtill greater ſhock to the conſtitution than an exceſs in fats? Is there any other article of food ſo hurtful as either, taken immoderately? Does it not

appear

appear evident, that an excefs at the end of a courfe of diet, is more hurtful than at the beginning of it?

17. I had a little head-ach laft night when I went to bed, was late in getting to fleep; pretty early in the morning had a foft ftool, ftill a little uneafinefs in my bowels, and not much appetite. Before breakfaft had a fmall liquid ftool, after which I was very uneafy in my bowels. I had, pretty frequently, wind from my ftomach, with now and then pricking pains in my bowels, and ineffectual attempts to go to ftool; no appetite for food, but was thirfty. Urine high-coloured. At bed-time was tolerably well. Is my prefent indifpofition owing in any meafure to the change of weather? I purpofed, after the honey-diet, to have tried fome of the fweet fruits, but I found every thing fweet fo difagreeable to me, that I rather chofe fomething extremely oppofite.

18. I flept pretty well, but, when I awoke in the morning, I felt much forenefs in my bowels, as if they had been bruifed, which made me figh and groan; this uneafinefs continued after I got up, and I had little or no appetite for breakfaft. The urine which I made yefterday was turbid. I felt univerfally ill, and oppreffed, with great uneafinefs in my bowels, and fometimes much noife in them. I paffed no wind downwards, but feveral times upwards. I was dull, very lazy, often fighed and moaned, and had no appetite for food. Four hours after rifing this morning I breakfafted on bread and infufion of rofemary, but had no appetite. Sufpecting that my prefent complaints might poffibly arife, in part, from the cheefe, I this day omitted it. In eating bread, I found the infide of

my

my mouth a little ſore. There were two or three ſmall pimples alſo at the corner of my mouth, and about as many large ones on my body. The uneaſineſs in my bowels, and univerſal diſtreſs, encreaſe when the hurricanes approach, and during their continuance I cannot ſtir, or even look up. Nothing paſſes through me, except ſometimes a little wind upwards, or downwards, and that without relief.

Here terminates Dr. Stark's Journal, with the affecting recital of his illneſs and ſufferings, during the laſt day of his life that he was capable of deſcribing them. The ſequel of this melancholy ſtory, with the account of the fatal cataſtrophe which ſoon followed, I ſhall defer until I have finiſhed with his other Experiments.

STATICAL

STATICAL EXPERIMENT,

O R,

OBSERVATIONS

Made on the Weight of the Body, with a View to determine how far it is affected, both in the Day and Night, by the Discharges of Perspiration and Urine.

THE daily food, during the time in which the following obſervations were made, was always (the 3d and 16th of December excepted) one pound eight ounces of flour, four pints of water, twelve drachms of ſalt, ſometimes with oil, of different kinds as marked in the Table, ſometimes without.

From the 5th to the 23d of December incluſive, the food was taken in equal portions, at two different times in the day; but both before and after this period, it was taken at three times. The quantity uſed at breakfaſt and dinner was nearly equal, and double what was uſed at night.

As the body was weighed every hour during the day, the waſte, or loſs of weight which it ſuſtained, from the inſenſible perſpiration and urine, was every hour exactly aſcertained; and the quantity of nocturnal perſpiration was, in like manner, eſtabliſhed, by weighing the body at going to bed, and immediately after riſing in the morning: and, by weighing it again directly after making water, the quantity or weight of the nocturnal urine was alſo known.

EXPLANATION

Of the ABBREVIATIONS, employed in

The following TABLE.

The Table is divided into columns, according to the day of the month. On one ſide are marked the hours after each meal, and directly oppoſite to them, the quantity of perſpiration, or of perſpiration and urine evacuated in each hour.

When two figures are joined by a crotchet, the oppoſite number marks the loſs of weight at the end of both hours.

The ſtate of the atmoſphere is marked by the letters ſ. c. r. f. m. (the initials of ſerene, cloudy, rainy, foggy, mixed,) placed immediately after the hour.

The letter w. placed immediately after a ſingle hour, ſignifies that I walked moderately in the open air, during all or moſt of the time; but when placed oppoſite to a crotchet it only implies that I walked part of the time.

b. ch. means ſitting in my bed-chamber.

ex. uſing moderate exerciſe in the houſe.

l. lying in bed.

ſl. aſleep in my chair by the fire.

n. ſitting quite naked by the fire.

b. buttoned up in my great coat.

d. at the further end of my dining-room, near the door.

Where there is no mark but what denotes the ſtate of the atmoſphere, it implies, that during that time I was at home, and nearly at reſt.

STATICAL TABLE.

November 29.	November 30.	December 1.	December 2.
6 oz. of suet, made into a pudding, with flour.	6 oz. of suet.	6 oz. suet.	4 oz. of suet.
Breakfast. Hour Perspiration oz. dr. 1 2 2 1 3 1 15 4, 5 } w. 4 9 Dinner. 1 2 7 2 1 15 3 1 12 4 1 14 5 1 8 9 lb. 1 2 1	After rising. Hour Perspiration oz. dr. 1 2 10 2 1 13 Breakfast. 1 3 2 2, 3 } w. 5 6 Dinner. 1 2 9 2 1 14 3 w. 0 11 4 2 7 Supper. 1 1 12 2 3 2 3 1 10 4 1 2 13 lb. 1 11 14	In bed 8 hours, in which time I perspired, 8 oz. 12 dr. After rising, Hour Perspiration oz. dr. 1 s. 2 8 2 s. 1 13 Breakfast. 1 s. 1 13 2 m. 2 0 3 s. 1 13 4 s. 1 11 Dinner. 1 s.w. 2 7 2 s. 2 8 3 s. 2 3 4 f. 2 7 Supper. 1 s. 1 2 2 s. 1 12 3 s. 1 6 13 lb. 1 9 7	In 8 hours nocturnal perspiration, 10 oz. 1 dr. After rising. Hour Perspiration oz. dr. 1 c. 2 11 2 c. 2 5 Breakfast. 1 m. 2 2 2 s. w. 1 6 3 s. 3 2 4 s. 1 13 Dinner. 1 s. 1 13 2 s. w. 1 14 3 s. 2 10 4 s. 2 1 Supper. 1 s. 1 8 2 s. 1 10 3 s. 1 5 4 s. 1 21 14 lb. 1 12 0

December

December 3.			4.			5.			6.		
No Food*.			4 oz. of suet.			4 oz. of suet.			4 oz. of suet.		
In 9 hours nocturnal perspiration, 10 oz. 4 dr.			In 8 hours, 30 minutes, nocturnal perspiration. 8 oz. 6 dr.			In 8 hours, 30 minutes nocturnal perspiration, 8 oz. 14 dr.			In 8 hours, 15 minutes nocturnal perspiration, 9 oz. 0 dr.		
After rising.			After rising.			After rising.			After rising.		
Hour	Perspiration oz.	dr.	Hour	Perspiration oz.	dr.	Hour	Perspiration oz.	dr.	Hour	Perspiration oz.	dr.
1	not observed		1 f.	2	15	1 c.	3	2	1 c.	3	2
2 c.	3	4	Breakfast.			2 c.	2	3	2 c.	·2	5
3 c.	1	13	1 c.	1	14	3 c.	2	1	Breakfast.		
4 c. l.	0	10	2 c.	1	12	4 c.	1	15	1 c. w.	3	6
5 c.	2	5	3 c. w.	1	11	5 c.	0	13	2 c.	2	7
6 c. l.	0	4	4 c.	2	11	6 c.	1	5	3 c. w.	3	3
7 c.	2	0	5 c.	2	2	Breakfast,			4 c.	2	12
8 c. l.	0	10	Dinner.			1 r. w.	3	5	Dinner.		
9 c.	2	1	1 c.	1	8	2 r.	2	5	1 c.	3	6
10 c. l.	0	9	2 c. w.	1	1	3 r. w.	1	6	2 c.	3	5
11 c. n.	1	11	3 c.	2	5	4 r.	2	8	3 c.	1	13
12 c.	1	6	4 c.	1	15	Dinner.			4 c.	2	1
13 c. n.	1	6	Supper.			1 c.	1	10	5, 6, 7 } c. w*.		
14 c.	1	2	1 c.	1	5	2 c.	1	13			
			2 c.	1	7	3 c.	1	9			
			3 c.	1	9	4 c.	1	3			
13 lb. 1	3	1	13 lb. 1	9	11	14 lb. 1	10	14	13 lb. 1	14	6

* I was induced to try the effect of long fasting, partly with a view to diminish the quantity of urine secreted in the night, which, from its copiousness, I found difficult to retain till the morning. From the 5th to the 8th hour after rising, I was very hungry, I then lost my appetite, became faint, weak, peevish, and, lastly, fell asleep.

During 8 hours of the day, viz. from the 3d to the 11th, I alternately sat by the fire in my dining-room, or in my bed-chamber, where there was no fire. During the last four hours, I alternately sat naked, or with my clothes on in the same place by the fire.

† On the 6th day, after dinner, I walked only in the beginning of the 5th hour, and towards the end of the 7th.

December

December 7.			8†.			9‡.			10.		
4 oz. of suet.			4 oz. of suet.			No oil.			No oil.		
In 8 hours nocturnal perspiration, 7 oz. 13 dr.			In 8 hours, 45 minutes nocturnal perspiration, 8 oz. 10 dr.			In 6 hours 15 minutes, nocturnal perspiration, 6 oz. 4 dr.			Nocturnal perspiration not observed.		
After rising.			After rising.			After rising.			After rising.		
Hour	Perspiration oz.	dr.	Hours	Perspiration oz.	dr.	Hour	Perspiration oz.	dr.	Hour	Perspiration oz.	dr.
1 c	3	1	1 f	2	11	1 f l	1	2	1 f l	1	1
2 m	1	12	2 f	2		2 f	3	10	2 f	2	3
3 m	2	9	3 f	1	12	3 f w	2	6	3 m	1	7
						4 f	2	3			
Breakfast.			Breakfast.			5 f w	1	7	Breakfast.		
1 f w	3	6	1 f	3	2	6 f	2	13	1 c ex	3	0
2 f	3	2	2 f w	2	6				2 c	2	2
3 f w	1	12	3 f	2	8	Breakfast.			3 r ex	2	7
4 f	*3	2	4 f w	1	8	1 f	2	3	4 r	1	12
			5 f	3	6	2 f	1	12	5 r ex	2	5
Dinner.						3 f	2	1	6	11	2
			Dinner.						Dinner.		
2 } 1 } f w	5	8	1 f b ch	1	5	Dinner.					
3 f	2	9	2 f	2	15	1 f d	1	10	1	1	8
4 f	2	5	3 f b ch	0	12	2 f	2	0	2 } 3 } fl	2	9
5	1	14	4 f	2	15	3 f d	1	2			
			5 f b ch	0	12	4 f	2	4	4	1	10
12 lb. 1	15	0	6	2	4	5 f d	0	14	5	1	9
						6 f	2	3			
			14 lb. 1	13	11				14 lb 1	9	5
						15 lb. 1	13	10			

* The unusual encrease in the quantity of the perspiration, during the hour immediately preceding dinner on this and the following day, was, I believe, owing to my having sat nearer the fire than I commonly do.

† This day, after dinner, I alternately sat in my bed-room, where there was no fire, or in my dining-room, where there was one.

‡ On the 9th and 10th, after getting up and weighing, I went to bed again, and lay for an hour without going to sleep. On the 9th, I, for 6 hours after dinner, alternately sat near the door of my dining-room, at a considerable distance from the fire, or at a moderate distance from it; and on the 10th, for 6 hours after breakfast, I alternately used moderate exercise, or sat still in my room.

December

December 11*.	12.	13.	14.
No oil.	No oil.	No oil.	4 oz. of ſuet.
Nocturnal perſpiration not obſerved.	In 7 hours, 20 minutes nocturnal perſpiration, 7 oz. 5 dr.	In 7 hours, 15 minutes nocturnal perſpiration, 6 oz. 15 dr. Urine collected in the night, 1 lb. 3 oz. 7 dr.	In 8 hours nocturnal perſpiration, 7 oz. 8 dr. Nocturnal urine, 1 lb. 9 oz. 12 dr.

December 11*. After riſing.

Hour		Perſpiration oz.	dr.
1	c	2	12
2	c	2	0
Breakfaſt.			
1	c	2	3
2	c	2	1
3	c	2	1
4	c	1	13
Dinner.			
1	c w	2	5
2	c	2	7
3	c w	2	7
4	c	2	1
5	r w	1	5
6	c	2	4
7	r w	1	8
8	r	1	14
14	lb. 1	13	1

12. After riſing.

Hour		Perſpiration oz.	dr.	Urine‖
1	ſ	2	8	
2	ſ	2	9	
Breakfaſt.				
1	ſ w	2	4	
2	ſ	3	10	
3	ſ w	1	7	
4	ſ	2	8	
5	ſ w	1	11	
Dinner.		Perſp.		Urine‖.
1	ſ	2	2	3 9
2		2	0	3 2
3		1	11	2 1
4		1	11	1 10
5		1	6	1 11
6		1	8	1 15
7		1	5	2 1
14 lb. 1		12	4	

In 7 hours urine, lb. 1 0 1

13. After riſing.

Hour		Perſp. oz.	dr.	Ur.	
1	f l	1	6	7	4
2	c	3	3	5	8
Breakfaſt.					
1	c	2	4	1	12
2	c	1	9	1	4
3	f	2	4	1	9
4	f	1	3	1	13
5	f r w	2	2	2	2
6	f	1	14	2	13
Dinner.					
1	f r w	1	6	2	8
2	c	1	14	1	13
3	m	1	12	1	12
4, 5, 6	m w	4	11	8	12
14 lb. 1		9	8	2 6	14

14. After riſing.

Hour		Perſp. oz.	dr.	Ur.	
1	c	2	9	3	1
Breakfaſt.					
1	r	2	5	1	0
2	r	2	4	0	9
3	r	2	7	0	13
4	r	1	6	1	8
5	c	1	4	1	5
6	m	1	12	1	11
Dinner.					
1	m w	2	0	1	11
2	c	1	15	1	12
3	m w	1	12	1	11
4		1	15	1	13
5		1	15	1	13
6		1	12	1	13
7		1	10	3	3
14 lb. 1		10	4	1 7	10

* During the 5th hour after dinner, I walked without my great coat; the reſt of the time, whilſt walking, I had it on.

‖ The quantity or weight of the urine ſecreted each hour, was determined by weighing the body immediately before and after making water.

December

December 15.	16.	17.	18.
4 oz. of fresh butter.	The food of this day mentioned below.	4 oz. of fresh butter.	4 oz. of fresh butter.
In 7 hours 45 minutes nocturnal perspiration, 7 oz. 9 dr. Nocturnal urine, 1 lb. 7 oz. 11 dr.	In 7 hours nocturnal perspiration, 6 oz. 12 dr. Nocturnal urine, 1lb. 1 oz. 7 dr.	In 7 hours nocturnal perspiration, 6 oz. 11 dr. Noctornal urine, 1 lb. 6 oz. 4 dr.	In 8 hours, 30 minutes nocturnal perspiration, 8 oz. 3 dr. Nocturnal urine, 1 lb. 11 oz. 7 dr.

December 15.

After rising Hour	Persp. oz.	dr.	Ur.	
1 c	2	11	2	8
2 c	2	1	1	15
Breakfast.				
1 c	1	12	0	13
2 c	1	11	1	1
3 c	1	15	1	7
4 m	1	15	1	12
5 c	1	9	1	7
6 c	1	14	1	12
Dinner.				
1 c w	1	15	1	13
2	1	7	1	0
3	2	1	2	0
4 m w	1	4	2	2
12 lb. 1	6	3	3 11	1

16.

After rising. Hour	Persp. oz.	dr.	Ur.	
1 c	2	6	2	8
2 c	2	1	1	6
Breakfast. Suet, yolks of eggs, of each 2 oz. water, 2 pints				
1 m	1	11	8	5
2 f	1	10	2	9
3 f	1	10	2	11
4 m	0	11	1	14
5 m	1	0	1	10
Dinner. Figs, 1 lb. Water, 2 pints.				
1 m	2	12	2	1
2 f	1	15	2	6
3 f w	2	2	*2	4
4 f	1	14	2	0
5	1	14	3	6
6	1	9	6	13
13 lb. 1	7	4	2 7	13

17.

After rising. Hour	Persp. oz.	dr.	Ur.	
1 f	1	3	1	2
2 c	2	13	1	2
Breakfast.				
1 f	1	3	0	3
2 f	1	8	0	11
3 f	1	14	1	5
4	2	8	1	1
5 f w	3	1	1	15
Dinner.				
1, 2 } fl	3	10	2	1
3	1	8	0	10
4 f w	2	5	1	7
5	1	13	1	14
6	1	4	1	11
13 lb. 1	9	1	0 15	2

18.

After rising. Hour	Persp. oz.	dr.	Ur.	
1 c	3	8	6	8
Breakfast.				
1 c	1	7	0	13
1, 3 } c	3	5	1	12
4 r	2	9	1	10
5 r	2	7	2	9
Dinner.				
1 c w	1	12	2	4
2	1	12	1	12
3	1	9	0	15
4 r w	1	10	0	14
5	1	12	1	1
6	1	5	0	14
7	1	7	2	0
13 lb. 1	8	7	1 6	10

* On the 15th I had two loose stools, and one on the 16th, immediately after breakfast.

December.

December 19.	20.	21.	22.
4 oz. of fresh butter.	4 oz. of fresh butter.	4 oz. of oil of marrow.	4 oz. of oil of marrow.
In 7 hours 15 minutes nocturnal perspiration, 7 oz. 8 dr. Nocturnal urine, 1 lb. 2 oz.	In 8 hours, 30 minutes nocturnal perspiration, 8 oz. 3 dr. Nocturnal urine, 1 lb. 2 oz. 7 dr.	In 8 hours 10 minutes nocturnal perspiration, 8 oz. Nocturnal urine, 1 lb. 8 oz. 14 dr.	In 8 hours, 15 minutes nocturnal perspiration, 8 oz. 5 dr. Nocturnal urine, 1 lb. 6 oz. 8 dr.

December 19.

Hour	Persp. oz. dr.	Ur.
After rising.		
1 ſ	2 8	2 7
Breakfast.		
1 ſ	2 1	1 3
2	1 11	0 15
3 c	1 13	1 0
4 c	1 13	1 0
5 m w	3 4	2 8
6 m	2 4	2 3
Dinner.		
1	1 12	1 0
2	1 6	1 0
3 c w	1 6	1 0
4	1 14	0 14
5	1 7	1 0
6	1 8	1 1
13 lb.	1 8 11	1 1 4

20.

Hour	Persp. oz. dr.	Ur.
After rising.		
1 c	3 5	2 5
Breakfast.		
1, 2 } c	3 10	0 15
3 m	1 8	1 13
4 m w	3 3	
5 m	1 14	2 2
6	1 14	2 1
Dinner.		
1	1 11	1 5
2	1 8	1 12
3, 4, 5 } ſ w	6 7	6 1
12 lb.	1 9 0	4 11

21.

Hour	Persp. oz dr.	Ur.
After rising.		
1 c	3 3	2 6
Breakfast.		
1, 2 } c	4 14	2 6
3 r w	1 10	1 8
4 c	2 6	1 4
5 c	2 1	1 6
Dinner.		
1 c w	2 8	1 8
2	2 6	1 4
3	1 13	1 12
4 ſ w	2 1	1 8
5	1 8	1 5
6	1 8	1 14
7	1 7	2 0
13 lb.	1 9 8	1 4 1

22.

Hour	Persp oz. dr.	Ur.
After rising.		
1 ſ	2 4	2 0
Breakfast.		
1 ſ	1 11	1 1
2 ſ	1 12	0 15
3 ſ	1 5	1 1
4 ſ	1 9	1 2
5 ſ	1 7	* 1 6
6		
7 ſ w	2 5	2 14
8	2 4	1 14
9	2 3	1 6
10 ſ w	0 10	1 9
11	1 12	0 11
Supper.		
1	1 14	0 12
2	1 7	0 10
13 lb.	1 7 7	1 5

* I had two loose stools on the 19th.

December

December 23.				
6 oz. of oil of marrow.				
In 8 hours nocturnal perspiration, 8 oz. 9 dr. Nocturnal urine, 14 oz. 11 dr.				
After rising.				
Hour	Persp. oz.	dr.	Ur. oz.	dr.
1 r	3	2	1	10
Breakfast.				
1 r	1	11	0	14
2 c	0	15	0	12
3 m	1	10	0	15
4 ſ	2	10	1	14
5 m w	2	1	2	13
6 m	2	1	2	8
7 ſ w	2	0	2	14
8	2	0	2	4
9	1	0	1	8
10 m w	1	4	1	5
11 ſl	1	15	0	13
Supper.				
1	1	11	0	11
2	1	3	0	13
14 lb.	1	9 3	1	5 10

24.				
6 oz. of oil of marrow.				
In 8 hours, 30 minutes nocturnal perspiration, 10 oz. 0 dr. Nocturnal urine, 15 oz. 12 dr.				
After rising.				
Hour	Persp. oz.	dr.	Ur. oz.	dr.
1 ſ	3	13	1	8
Breakfast.				
1 ſ	2	2	1	8
2 ſ	1	11	1	6
3 ſ	1	11	1	3
4 m w	2	8	2	8
5 ſ w	2	11	3	12
Dinner.				
1 b	1	10	1	8
2	1	9	2	0
3 b	1	8	1	8
4	1	10	1	12
5 ſ w	2	8	2	8
6	1	13	2	12
12 lb.	1	9 12	1	7 13

25.				
6 oz. of oil of marrow				
In 8 hours nocturnal perspiration, 9 oz. 4 dr. Nocturnal urine, 1 lb. 9 oz. 10 dr.				
After rising.				
Hour	Persp. oz.	dr.	Ur. oz.	dr.
1 r	2	15	2	3
Breakfast.				
1 r	1	15	1	11
2 r	1	9	1	6
3 r	1	13	1	3
4 c	1	7	1	10
5 c	1	5	1	7
Dinner.				
1 c w	2	3	2	4
2 r w	2	12	2	0
3 c w	2	8	1	8
4	2	7	1	10
Supper.				
1 ſl	1	13	1	2
2, 3 } ſl	2	12	1	14
13 lb.	1	10 14	1	3 14

26.				
In 7 hours, 30 minutes nocturnal perspiration, 8 oz. 2 dr. Nocturnal urine, 1 lb. 5 oz. 8 dr.				
After rising.				
Hour	Persp. oz.	dr.	Ur. oz.	dr.
1 c*	2	14	2	12

* Having this day been very much indiſpoſed, I was obliged, at preſent, to diſcontinue my obſervations.

By adding together the particular numbers contained in this Table it appears,

That in 355 hours, during the day-time, the perſpiration was, 698 oz. 7 dr. And

That in 190 hours 15 minutes, during the night-time, it was, - 196 oz. 14 dr.

And by adding together the particular numbers, in that part of the Table ſubſequent to the 12th of December it appears,

That in 169 hours, during the day-time, the perſpiration was, - 324 oz. 2 dr.
The urine was, - - - - 300 oz. 6 dr.
That in 109 hours 40 minutes, during the night, the perſpiration was, 111 oz. 9 dr.
The urine was, - - - - 297 oz. 6 dr.

And hence, by a ſhort calculation it will be found, that the hourly waſte of my body was nearly equal, both day and night, being about 3 oz. 10 or 11 dr.

The influence of the food upon the perſpiration and urine may, in ſome meaſure, (though I own imperfectly) be judged of, from the following Table.

Diet

Diet.	Day*.	Perſpiration.		Urine.		Night†.		Perſpiration.		Urine.	
	Hours.	oz.	dr.	oz.	dr.	Hours.	Min.	oz.	dr.	oz.	dr.
No food	13	19	1	——		8	13	8	6	——	
No flour	13	23	4	39	13	7		6	11	22	4
No oil	71	137	12	——		22	35	21	12	——	
No oil	21	——		50	15	15	15	——		45	3
4 oz. marrow	26	48		37	6	16	15	16	14	37	3
4 oz. butter	63	121	6	95	6	39	25	38	10	106	3
4 oz. ſuet	80	160	14	——		47	30	48	2	——	
4 oz. ſuet	14	——		23	10	7	45	——		23	11
6 oz. marrow	39	77	13	65	5	24		27	6	4	14
6 oz. ſuet	49	99	6	——		25		29	1	——	

* That is, whilſt out of bed.

† That is, whilſt in bed.

A Continuation of the STATICAL TABLE.

February 5.			6.			7.			8.		
Heated honey, 8 oz.			Honey, 8 oz.			Heated honey, 12 oz.			Honey, 6 oz.		
Breakfast.			Dinner.			Breakfast.			Breakfast.		
Hour after.	Perſp. oz. dr.	Ur. oz. dr.	Hour after.	Perſp. oz. dr.	Ur. oz. dr.	Hour after.	Perſp. oz. dr.	Ur. oz. dr.	Hour after.	Perſp. oz. dr.	Ur. oz. dr.
1 } ſ w*	5 12	10 9	1 ſ	2 3	1 10	2 ſ	3 0	2 10	1 ſ	2 6	1 3
2 } m			2 ſ	2 4	1 8	3 ſ	2 5	5 9	2 ſ	2 0	1 1
3 m ſ	3 0	7 10	3 } w			4 ſ	2 0	7 8	3 ſ	1 15	1 0
4 m ſ	2 8	5 6	4 }						4 ſ	1 15	0 15
Dinner.			Supper.			Dinner.			Dinner.		
1 c ſ	1 14	7 6	1 w	2 0	1 14	1 ſ	2 0	7 10	1 } w‡	10 14	4 15
2 c w	2 0	10 10	2 ſ	2 1	1 9	2 } w†	6 0	10 4	2 }		
3 c ſ	2 12	5 4	3 ſ	1 9	1 6	3 }			3 }		
									4 }		
									Supper.		
									1 } ſ	3 14	2 0
									2 }		

* Walked all the time till in a breathing ſweat.

† Walked briſkly all the time in a cold wind.

‡ Walked about half the time.

February

February 9.			10.			11.			13.		
No Honey.			Heated honey, 4 oz.			Honey, 4 oz.					
Breakfaſt.			Breakfaſt.			Breakfaſt.			Dinner.		
Hour after.	Perſp. oz. dr.	Ur. oz. dr.	Hour after.	Perſp. oz. dr.	Ur. oz. dr.	Hour after.	Perſp. oz. dr.	Ur. oz. dr.	Hour after.	Perſp. oz. dr.	Ur. oz. dr.
1 ſ*	2 14	0 14	1 c ſ†	4 0	1 11	1 c ſ	2 10	2 4	1 c ſ	2 9	5 10
2 } 3 } ſ	4 2	2 1	2 c ſ	2 3	1 13	2 c ſ	2 0	1 14			
			3 ſ	1 12	1 10	3 c w	1 4	2 4			
4 ſ w	3 8	1 4	4 c w	2 0	1 12				Feb. 16.		
Dinner.			Dinner.			Dinner.			Unwell.		
1 ſ w	3 6	2 0	1 } 2 } w	3 15	3 12	1 c ſ	3 1	2 3	Supper, mulled Port wine‡ and bread.		
2 c ſ	2 14	1 6	3 ſ	3 1	4 8	2 c w	1 2	6 2			
						3 ſ	2 7	3 7			
Supper.						Feb. 12.			1 ſ	1 13	1 12
1 ſ	2 2	0 14				Heated honey 4 oz.			2 ſ	1 14	2 0
						Breakfaſt.			Head-ach.		
						1 c ſ	2 13	2 2	Feb. 18.		
						2 c w	1 15	2 8	After riſing.		
						3 c ſ	2 9	2 7	3 m ſ	2 3	1 2
									4 m ſ	1 13	0 14

* Sitting by the fire, or ſtanding in my bed-room.

† Sitting by the fire, excepting for a little time, in my bed-room, whilſt I bathed my feet.

‡ My urine was increaſed in quantity after drinking the Port wine.

It

N. B. It ſhould be remarked that the perſpiration, during the night-time, and in the morning, before breakfaſt, is influenced by the food taken the preceding day, and therefore, ſtrictly ſpeaking, belongs to it, though their place in the Table muſt, neceſſarily, ſtand as it does.

Several other remarks might be made on the Table, but they will probably occur to the reader himſelf from the peruſal of it. I ſhall therefore only add, that although I have been extremely careful to avoid miſtakes, yet I am ready to confeſs, that wherever any uncommon encreaſe or decreaſe in the weight of the body is obſerved, it is more probable that I ſhould have been miſtaken, than that any thing uncommon ſhould have happened. I have likewiſe to beg of the Reader to remember, that theſe Obſervations were made, not ſo much in hopes of determining any thing on this ſubject, as of diſcovering how the land lay, and of enabling me to undertake ſome more accurate and deciſive Experiments.

A N

A C C O U N T

O F

Dr. STARK's laſt ILLNESS and DEATH,

B Y

Sir JOHN PRINGLE, or by Dr. SAUNDERS, moſt probably the former*.

Quis talia fando————
temperet a lacrymis?

DR. STARK died in the twenty-ninth year of his age. He was of a fair complexion, tall, of a thin make, and healthful. For ſeveral months before his death he had been employed in making experiments upon himſelf, of the effects of different kinds of food; among the laſt was that of honey and flour made into a pudding, upon which he had lived ſeveral days; and which ſeemed to

* *For this account I am indebted to Dr. Garthſhore, who tranſcribed it ſome years ago from the original copy, in the poſſeſſion of Dr. Huck Saunders.*

to be extremely diuretic at firſt*, as he made conſiderably more water than the liquor he drank. At laſt it brought on a diarrhœa, for which he ate Cheſhire cheeſe, to the quantity of a quarter of a pound, without any other food, and that ſeemed to bind his body ſo much that he had not been at ſtool for five days†. When he was taken ill, on Sunday, the 18th of February, 1770, he ſent for Mr. Hewſon to bleed him, when he complained of his head and in his belly, The blood was ſomewhat ſizy. He had uſed ſome opening medicines without effect, until the 20th, that he took the Ol. Ricini, which procured five or ſix motions. On the morning of the 20th, he complained of an oppreſſion and ſickneſs at his ſtomach, and he had ſpit ſome blood in the night; his pulſe was very quick, and he had other feveriſh ſymptoms. He had no ſleep for two nights, nor did he ſhut his eyes afterwards. Ordered tartar. emetic. gr. v. ſal. rupell. ℥ ſs to be diſſolved in a pint and a half of water, and of this, a coffee-cup full every ten minutes, till it had a ſenſible operation. This was directed, upon the ſuppoſition that he had ſome load in his ſtomach and bowels, which was to be relieved by vomiting and purging. He took three cupfulls, in all, of the medicine, vomited thrice, and had ſeven looſe ſtools, but complained of great ſickneſs

* The diuriſm he aſcribed to the boiling of honey, not having obſerved that quality of it when uſed in its natural ſtate.

† That he had eaten nothing but Cheſhire Cheeſe is not certain, it was at leaſt two days, one of the Gentlemen who attended him thinks more.

It appears from Dr. Stark's own Journal, that the two preceding remarks are not perfectly correct.

ſickneſs and lowneſs after them. The next day, (21ſt) he was extremely low, had the anxietas præcordiorum in a great degree, reſtleſſneſs, fluſhings in his cheeks, and complained much of a great flow of ſweet ſaliva in his mouth, which made him ſick. The looſeneſs ſtill continued. The following mixture was directed, ℞ julep. e creta. ℥ vii ſs, tinct. cinnamon. ℥ ſs, tinct. thebaic. gutt. x. m dentur coch. iv. poſt alternas ſedes liquidas. Of this he took one doſe, and had no ſtool after it. At this time he ſeemed much worſe, he ſpoke ſlow and low, and ſeemed with difficulty to recollect, or pronounce the word he wanted to utter.

During the night he was very feveriſh, and ſo delirious as to attempt getting out of bed. The purging returned, and the ſtools were bloody and involuntary. He ſometimes coughed and brought up ſome mucus, tinged with blood. A bliſter, which had been applied the night before, roſe well, but without any other effect. A decoction of the bark and camomile-flowers, with ſome Port-wine, was thrice injected as a clyſter, which ſtopped his purging. He continued to grow worſe, and died on Friday, the 23d.

Here follows Mr. Hewſon's Account of the Illneſs and Inſpection of the Body; which is added, as he was more with the Patient than any of the Phyſicians who attended him.

On Sunday, the 18th of February, Dr. Stark ſent to deſire me to bleed him; I went at nine, and found him going to take

 a clyſter.

a clyſter. He told me he had pain in the lower part of his belly, that he had not made water in any quantity, nor had had a ſtool for three or four days; this he attributed to a change in his diet, *viz.* from a pudding, made of honey and flour, to cheeſe, of which I underſtood he had eaten to the quantity of three or four pounds, without having had any evacuation ſince he began it, and this, he told me was the oppoſite effect to that of honey, for, whilſt living on the pudding of flour and honey, he had made more urine than he had drank water, which was all his drink. Agreeable to his deſire I took away nine ounces of blood, which was received into four cups; the two firſt had an inflammatory cruſt. The blood, at five o'clock, P. M. had very little ſerum, which I aſcribed to its having ſtood in a cool place, as the coagulum felt very firm, and as one cup, which was removed into a warm room, had more ſerum ſeparated the day following. Soon after the bleeding he took half an ounce of caſtor oil. In the afternoon he thought himſelf rather better, having made water, and diſcharged ſome fæces, which he told me were extremely offenſive. Upon enquiring whether he had been ſenſible of any enlargement of his bladder, he anſwered in the negative, and obſerved that obſtructions there had not been total, for that he had frequently made a ſpoonful of water, and could, at any time, diſcharge a ſmall quantity. He drank, during the day, plentifully of water-gruel, with a little juice of orange in it.

On Wedneſday morning, I found that he had been very reſtleſs, hot, feveriſh and thirſty, throughout the preceding night. He ſaid that he had ſpit blood, and complained of a pain in his head. His face was remarkably florid, he ſeemed much oppreſſed, and fetched his breath every now and then with a moan. His

ſkin

ſkin was very hot, his pulſe ſeemed to require another bleeding, which he deſired me to perform, but hearing that he had ſent for a Phyſician, another medical friend, I deſired he would defer the operation till after his viſit. I returned at twelve, and underſtood that he had been deſired to repeat his caſtor oil, but not to bleed. Upon examining his pulſe, I was ſurprized to find it ſo much altered in ſo ſhort a time, for it was remarkably ſoft, and it was upon this change it was thought improper to open a vein. I ſaw him again at five o'clock in the afternoon, and found him much oppreſſed; he moaned frequently, ſaid his ſtomach loathed every kind of watery liquor, complained of a violent pain in his forehead, was very low-ſpirited, and told me he apprehended he ſhould not outlive the night. That evening he was directed to uſe the following medicine, tart. emetic. gr. v. ſal rupell ℥ ß diſſolved in a pint and half of water, of this he took, at intervals, about a third part by cupfulls, till it operated. I ſaw him about an hour afterwards, and he thought himſelf much relieved, though nothing had come up from his ſtomach, but the water he had drank and a little mucus.

On Wedneſday I found him very low, and conſtantly ſpitting. He told me his ſaliva was ſweet, and ſuppoſed that his purging was owing to his ſwallowing it in his ſleep, for that when he ſpit it out he ſeemed to purge leſs. The pain in his belly, he ſaid, ſeemed not ſo low down as it had been, but added, that the pain of his head was intolerable. He took, during the day, ſome chalk julep, with two or three drops of laudanum, after every ſtool. In the evening he told me he was afraid to go to ſleep, leſt he ſhould ſwallow his ſaliva.

On Thurſday-morning I underſtood he had been delirious in the night, and had got out of bed in ſpite of his nurſe, but had immediately tumbled down on the floor. When I ſaw him he muttered his words ſo that I could not underſtand him, but ſeemed ſenſible of what I ſaid to him, and gave me his hand to feel his pulſe. At two in the afternoon I found him evidently worſe, for he was then inſenſible, and his ſtools were frequent and involuntary, and, as the nurſe expreſſed it, nothing but diſcoloured water. He was bliſtered, took glyſters of a decoction of the bark, and uſed the julepum vitæ of Bates for a cordial, but from this time he grew worſe and worſe, and died next day at four in the afternoon.

The body was examined by Mr. Hunter and myſelf, on Sunday, at one o'clock. Upon opening the abdomen two or three ounces of water were found in the pelvis. The bladder contained about ſix ounces of urine, of a natural colour. The ſmall inteſtines appeared very red and inflamed at particular parts, which, upon opening into their cavities, was found to be the glandulæ peyerianæ enlarged. One cluſter of theſe ſeemed ulcerated. Some of the glandulæ ſolariæ were of the ſize of a ſplit pea. The meſenteric glands were likewiſe enlarged, and, when cut into, were found to be remarkably ſoft and tender. The ſtomach, near its upper orifice, internally, had the veſſels of its villous coat tinged with blood which burſt* on a very ſlight preſſure. The liver ſeemed rather ſmall. The ſpleen rather larger than common, but had no morbid appearance. The kidnies had their veins fuller of blood than uſual, but the ureters and pelvis were of a natural ſize.

* The expreſſion is *broke down*, in the original.

ſize. The larger inteſtines ſeemed quite ſound. In the thorax there was found more water than even in people who die a violent death, even after lying two days before diſſection. The ſame was obſerved of the pericardium. The lungs had ſeveral black ſpots in different parts of their ſubſtance, owing to extravaſated blood. The heart ſeemed flaccid, and had no coagulum in it, the blood being fluid; however, one or two tranſparent coagula were afterwards found in the veſſels of the brain, but they were very ſoft.

The dura mater had no morbid appearance; but the veſſels of the pia mater had more moiſture in the cellular membrane, contiguous to them, than is natural. The ventricles contained each about a tea-ſpoonfull of water, and that in the left was of a bright yellow colour. The pineal gland had ſeveral earthy particles in it. The other parts of the brain had no preternatural appearance.

This was Mr. HEWSON's *written Account of the Diſſection.*

Mr. HUNTER *gave the Phyſician the following Account of the Appearance ſome days afterwards, from his Memory.*

The brain had no morbid appearance, except that in the left ventricle; the ſerum, which was not more than uſual in quantity, had a ſlight bloody caſt. The ſubſtance of the brain was of a natural

natural firmneſs. In the thorax the lungs had a ſlight adheſion to one ſide, and there were maculæ, ſome of them as broad as a ſhilling, all over the ſurface of that organ; owing to an extravaſation of blood in the cellular membrane, and under the common membrane of the lungs. In the ſubſtance of the lungs the cellular membrane contained a good deal of extravaſated blood. In the cavity of the thorax there was more than the natural quantity of ſerum. The heart was ſound, but upon opening it and the great blood-veſſels, the blood was found in a reſolved ſtate, that is, about the conſiſtence of ſyrup without any polypus concretion or coagulation. The liver was ſound. The gall-bladder was half full of bile, and of a natural colour. Nothing extraordinary was contained in the ſtomach and inteſtines. There were no marks of inflammation on the ſtomach, but there were on the inteſtines, eſpecially towards the lower end of the ileum, where the peyerian glands were found enlarged beyond their natural ſize, in ſo much that they could be felt with the fingers on the outſide of the gut. There was no extravaſated blood in any part of the tube.

Mr. Hunter took notice in this ſubject, of the beginning diſſolution of the internal coats, near the great end of the ſtomach, but which he accounted no morbid appearance, as it had been obſerved on other occaſions.

After giving the above account Mr. Hunter added, that he had forgot to mention the diſeaſed appearance he had obſerved in the meſenterick glands. They were larger than common, and when cut into were obſerved to be much paler than natural, and their ſubſtance to be ſo ſoft as to appear like a pulp.

F I N I S.

INDEX.

A

B

C

D

INDEX.

Morbid

INDEX.

P

R

S

T

I N D E X.

EXPLICATION

EXPLICATION
OF THE
FIGURES.

FIGURE I. PLATE I.

Repreſents a portion of the higher part of the colon, taken out of the body of the man, (Part I. Ch. 1. § 3. p. 4.) and inverted.

a. A broad eroſion of the internal coat.
b, &c. Smaller eroſions of the ſame coat.
c, &c. Small black ſpots ſhining through that coat.

FIGURE II. PLATE II.

Repreſents the internal ſurface of the rectum, and that of the adjoining part of the colon, taken out of the body of the woman, (Part I. Ch. 1. § 4. p. 5.) and cut open.

A B. The circulus albus, and the boundary between.
C. The ſkin, and
D. The internal coat.

Above the circle appear the ſinus ſurſum cavi, deſcribed by Haller (prim. lin. DCCXLII.) A great portion of the lower part of the rectum, being quite ſound, is folded up.

a a. Hemiſpheres filled with a gelatinous ſubſtance.
b, &c. Veſicles of the internal coat, out of which the gelatinous ſubſtance having been expreſſed, blown up with air, and having one, two or three openings, into ſome of which a hog's briſtle is introduced.
c. Openings of veſicles not blown up.
d. Large irregular openings in the internal coat.
e e. Black ſpots appearing through that coat.
f. A warty excreſcence.

FIGURE III. PLATE II.

Repreſents the internal ſurface of the middle portion of the colon, taken out of the ſame body, and cut open.

A. The middle point of the large inteſtines.

B. The

B. The ſuperior extremity of this portion.

C. The inferior extremity of the ſame.

D. Two lymphatic glands.

a a, &c. Irregular eminences of the internal ſurface, which, towards the upper extremity were placed in two parallel lines, between which was a very long livid depreſſion.

b b. Hemiſpheres filled with a gelatinous ſubſtance, each having a pellucid middle point, at which, in one of the hemiſpheres, a hog's briſtle is made to paſs a little way.

c, &c. Veſicles, emptied of the gelatinous ſubſtance they had contained, blown up, having each two openings, through which a hog's briſtle is made to paſs.

d. The orifices of veſicles not blown up.

e, &c. Irregular eroſions of the internal coat, and ſometimes of the cellular ſubſtance.

FIGURE IV. PLATE II.

Repreſents the internal ſurface of a portion of the lower part of the rectum, taken out of the body of the man (Part I. Ch. 1. § 4. p. 5.)

In the middle part of the figure is repreſented a large hemiſphere, which, before the portion of inteſtine had been put into ſpirits, was much fuller than it here appears to be, and it was in ſome degree tranſparent.

FIGURE V. PLATE I.

Repreſents the internal ſurface of the adjoining portions of the ileum and of the colon, taken out of the body of the woman (Part I. Ch. 1. § 4. p. 5.) and cut open:

A. The Appendix vermiformis.

A B. The lower portion of the ileum, on which appear numerous roundiſh eminences, becoming gradually ſmaller towards the higher part, where they almoſt diſappear.

A C. The higher portion of the colon.

D. The valve of the colon cut open.

a, &c. Large veſicles with one or more orifices.

b. Two irregular openings in the internal coat, probably the baſes of veſicles; one of them is a little raiſed up by a briſtle.

c. Two large eroſions.

FIGURE

EXPLICATION OF THE FIGURES.

FIGURE VI. PLATE I.

Reprefents the internal furface of a fmall portion of the lower part of the colon, taken out of the body of a woman who had laboured under a bad purging for four months before her death.

On it appear many very fmall hemifpheres.

FIGURE VII. PLATE I.

Reprefents a portion of the lower part of the ileum, taken out of the body (Part. I. Ch. 1. § 5. p. 7.) the glands eroded in three places.

FIGURE VIII PLATE I.

Reprefents the internal furface of a portion of the lower part of the ileum, taken out of the fame body.

A B. A longitudinal eminence, formed by a fmall remaining portion of the mefentery pufhing up part of the inteftine.

a, &c. Small holes in the valvular conniventes.

b, &c. Holes in the part of the internal coat, neareft the mefentery.

c. A portion of the internal coat, furrounding a hole, raifed up by air blown into the cellular fubftance.

FIGURE IX. PLATE III.

Reprefents the femi-lunar valves of the aorta, convex towards the ventricle, and almoft fhutting up the paffage, as they appeared on drawing afide the large portion of the tricufpidal valve.

A B C. The three femi-lunar valves.

D. The large portion of the tricufpidal valve drawn afide.

E F G. The internal furface of the left ventricle.

H. The feptum cordis.

FIGURE X. PLATE III.

Reprefents the three femi-lunar valves, with the neighbouring parts of the left ventricle and of the aorta, laid fully open to view.

A. The cavity of the aorta.

b. The

B. The orifice of the right coronary artery.

C. The orifice of the left coronary artery.

D E. The internal ſurface of the left ventricle.

F. Part of the large portion of the tricuſpidal valve.

G H K. The ſemi-lunar valves ſtanding at a diſtance from the ſurface of the aorta, and partly covered with fatty excreſcences.

H. One of the valves cut up to ſhew the increaſe in thickneſs, which is chiefly at the lower part, and appears better at

L. A ſmall portion of the valve k.

FIGURE XI. PLATE III.

Repreſents the external ſurface of a portion of the dura mater, taken from the upper and anterior part of that membrane, and out of the body of the woman. (Part IV. Ch. I. § 2. p. 70.)

A B. A hollow formed by the upper ſide of the longitudinal ſinus, ſinking down between two eminences, occaſioned by the two lower ſides of that ſinus being, after they had been cut aſunder, drawn aſide. Near that hollow, and on either ſide of it, is repreſented the uneven ſurface of a diſeaſed portion of the dura mater; that ſurface was not white or ſhining, but of a dark aſh colour, and moiſtened with pus; the boundaries of it were in ſome parts quite black.

C D. Two portions of the external lamina of the dura mater raiſed up by blowing, into the form of bliſters. In each appear ſeveral apertures, at which, on preſſing the neighbouring parts, pus had iſſued. At an aperture in each bliſter, a briſtle is made to enter, both of which, as air had before done, found a paſſage between two laminæ.

E. A thin portion of the external lamina, puſhed up by one of the briſtles, which ſhines through it.

F. The extremity of the other briſtle, paſſing out at an opening, (through which both pus and air had paſſed) on the outſide of one of the superior angles of the longitudinal ſinus, which is here cut acroſs. The briſtle, as it paſſes along, is repreſented ſhining through ſeveral thin portions of the external lamina.

Neither matter, nor air, nor either of the briſtles, found any paſſage through the internal lamina, which did not appear in one part thinner than in another, or into the longitudinal ſinus.

A CATAL.

BOOKS in ANATOMY, MEDICINE, SURGERY, and NATURAL PHILOSOPHY, printed for J. JOHNSON, No. 72, St. Paul's Church-Yard.

ANATOMICAL PLATES.

1 A COLLECTION of Engravings, tending to illustrate the Generation and Parturition of Animals and of the Human Species, by Thomas Denman, M.D. Licentiate in Midwifery of the College of Physicians. Price 10s. 6d. in Boards, in large Quarto.

A second Number of this Work, which will contain twelve Plates, is in considerable forwardness.

2 Dr. Hunter's superb Plates of the Gravid Uterus, as large as life, with Explanations in Latin and English, 3l 13s 6d

3 Forty Anatomical Tables of the Gravid Uterus, as large as Life; with Explanations, and an Abridgement of the Practice of Midwifery, by William Smellie, M.D. price 2l 5s in boards, in royal folio

A new Edition, carefully revised and corrected; with Notes and Illustrations, adapted to the present improved Method of Practice. By A. Hamilton, M. D. F. R. S. Edin. and Professor of Midwifery in the University of Edinburgh.

*** The Writers of the Monthly Review for September last, speaking of the small Copies of these Plates, very justly observe, that by reducing the Figures, so as to bring them into an octavo size, the original intention of the Author is frustrated. It is on a perfect knowledge of the size and proportion of the bones that the whole practice depends; and, on that account, by exhibiting the figures in their full natural size and position, Dr. Smellie's plates ever have been, and most probably ever will be, the best means of conveying a proper idea of the parts to such students as have not the opportunity of a long attendance at an Anatomical Theatre; and that the price is a small sum compared to the usefulness of the work.

4 Albinus's Anatomical Figures of the Human Body; with a Supplement, containing the Blood Vessels and Nerves, Royal Folio. Engraved on Fifty-one large Copper Plates, Fifteen Inches by Twenty-two. With Explanations in a separate Volume. Price 3l. 3s. in Boards.

BOOKS and PAMPHLETS.

5 Aikin on the Use and various Preparations of Lead, with Remarks on Goulard, 1s 6d
6 ——— Thoughts on Hospitals, 1s 6d
7 ——— Translation of Beaumé's Manual of Chemistry, with Notes, 3s 6d sewed,
8 ——— Œconomiæ Animalis delineatio; in usum Juventutis Medicæ, 1s 6d
9 ——— Sketch of the Animal Œconomy, 1s 6d
10 ——— Edition of Dr. Lewis's Materia Medica, with Additions, 4to, 1l 4s boards
11 ——— Biographical Memoirs of Medicine in Great Britain, from the Revival of Literature to the Time of Harvey, 4s sewed
12 Astruc on the Venereal Disease, 4to, 10s 6d in boards
13 Alanson on Amputation, 2d edit. greatly enlarged, 6s bound
14 Alexander's Experimental Enquiry concerning the Causes which have generally been said to produce putrid Diseases, 3s 6d sewed
15 ——— Experiments on Antiseptics in putrid Diseases; on the Doses and Effects of Medicines; and on Diuretics and Sudorifics, 3s 6d sewed
16 Black's Observations on the Small-Pox and Inoculation, sewed 3s 6d
17 ——— Historical Sketch of Medicine and Surgery, from their Origin to the present Time; and of the principal Authors, Discoveries, Improvements, Imperfections and Errors, 6s
18 Butter's Account of Puerperal Fevers, illustrated by Cases, 2s 6d sewed
19 ——— on the Infantile Remittent Fever, commonly called the worm Fever, 1s
20 ——— on opening the Temporal Artery, 4s sewed
21 Buckner's easy Method to make Deaf Persons Hear, 1s 6d
22 Bell on Ulcers, 8vo, 6s boards
23 ——— System of Surgery, 5 vols, 1l 10s boards
24 Brooke's Practice of Physic, 2 vols, 10s bound
25 Crawford on Animal Heat, a new edition, very much enlarged, 8vo, 7s in boards

26 Cullen's First Lines of the Practice of Physic, 4 vols, 1l. 4s boards
27 Cullen's Nosologia Methodica, 2 vols, 12s in boards
28 Curry on the Nature of Fevers; on the Cause of their becoming so frequently Mortal, and on the Means to prevent it, 1s 6d
29 Dawson's Cases in the Acute Rheumatism and Gout, with the Method of Treatment, 6s
30 ——— Account of a safe and efficacious Medicine in Sore Eyes and Eye-lids, 1s
31 Denman on Puerperal Fevers, 1s 6d
32 ——— Natural Labour, 2s
33 ——— Preternatural Labour, 2s
34 ——— Uterine Hemorrhages, 2s
35 ——— Difficult Labour, Part I. 2s
36 Edinburgh New Dispensatory, 6s 6d boards
37 Else on the Hydrocele, and other Works; with some new Cases of the Hydrocele, by G. Vaux, 2s 6d sewed
38 Elliot's Physiological Observations 1s 6d
39 ——— Medical Pocket-Book; containing a short Account of the Symptoms, Causes, and Methods of Cure, of the Diseases incident to the Human Body, 2s. sewed
The same printed on one Side only, the other left blank for additional Receipts, 2s 6d
40 ——— Account of the Medicinal Virtues of the principal Mineral Waters in Great Britain and Ireland, and on the Continent, with the Method of impregnating Water with fixed Air, invented by Dr. Priestley, and improved by others, 3s sewed
41 ——— Elements of the Branches of Natural Philosophy connected with Medicine, viz. Chemistry, Optics, Sound, Hydrostatics, Electricity and Physiology. With Bergman's Tables of Elective Attractions, 5s sewed
42 Fothergill's Works, with Memoirs of his Life, by Dr. Elliot, 6s sewed
43 Fordyce's (George) Practice of Physic, 5th edit. 5s. sewed
44 ——— Elements of Agriculture, a Syllabus of his Chemical Lectures, 2s 6d sewed
45 Falconer's Observations and Experiments on the Poison of Copper, 2s. sewed,—N. B. This may be had, bound up with Dr. Percival on the Poison of Lead, price 4s
46 Fourcroy's Elements of Chemistry, 4 vols, 8vo, 1l 4s, boards
47 Report of Dr. B. Franklin and other Commissioners, charged by the King of France with the Examination of Animal Magnetism, as now practised at Paris by Mesmer and others for the Cure of various and obstinate Diseases. 2s 6d
48 Gregory's Conspectus Medicinæ, 2 vols, 13s boards
49 Hossack's Abridgement of Van Swieten, 5 vols, 1l 10s
50 Hunter's (W.) Medical Commentaries, 6s boards
51 ——— Two introductory Lectures to his general Course of Anatomical Lectures, 4to, 6s
52 Hunter's (John) Natural History, Anatomy, and Diseases of the Human Teeth, with 16 Copper-Plates, 4to, 1l 1s bound
53 ——— on the Venereal Disease, 4to, 2d edit. 1l 1s in boards
54 ——— on the Animal Oeconomy, 4to, 16s in boards
55 Hewson's Experimental Enquiries, on the Blood, 3s sewed
56 Hoffman's Practice of Physic, with a great Number of Cases, translated by Dr. William Lewis, and revised by Dr. Duncan, 2 vols, 8vo, 12s in boards
57 Henry's Experiments on the Preparation, Calcination, and Medicinal Uses of Magnesia Alba: On the Solvent Powders of Quick Lime: On Absorbents: On the Antiseptic Powers of Vegetable Infusions prepared with Lime, &c. On the sweetening Properties of fixed Air, &c. 2s 6d
58 Henry's Medicinal Virtues of Magnesia, 6d
59 ——— Method of preserving Water at Sea, 2s
60 ——— Memoirs of the Life of Haller, 2s 6d sewed
61 Halleri Primæ Lineæ Physioligæ, 7s, bound
62 Hamilton's Midwifery, 5s boards
63 Huxham on Fevers, 5s bound
64 Haygarth's Inquiry how to prevent the Small-Pox: And Proceedings of a Society for promoting general Inoculation at stated Periods, and preventing the Natural Small-Pox in Chester, 3s in boards

65 Innes's Defcription of the Mufcles, 2s 6d fewed
66 ——— Anatomical Tables, 6s fewed
67 Johnfon's New Syftem of Midwifery, founded on Practical Obfervations, 4to, with Plates, 1l 2s in boards
68 ——— Cautions to the Heads of Families: containing Directions to Nurfes who attend the Sick, and Women in Child-bed, 2s 6d
69 Kirkland's Inquiry into the prefent State of Medical Surgery, 2 vols 12s 6d boards
70 London Medical Journal. N. B. This Work is continued in Numbers, Quarterly, price 1s 6d each
71 Lavoifiere's Effays: containing a Hiftory of Difcoveries relating to Air, &c. with original Experiments, tranflated from the French by T. Henry, F. R. S. with Notes, 7s.
72 ——— on Atmofpheric Air, with a particular View to inveftigate the Conftitution of the Acids, tranflated by the fame, 2s 6d fewed
73 Le Dran's Operations in Surgery, 7s
74 ——— Obfervations, 5s
75 Lewis's Materia Medica, a new edit. with Additions, by J. Aikin, price 1l 4s in boards
76 ——— New Difpenfatory, 8vo, 7s bound
77 ——— Abridgement of the Edinburgh Medical Effays, 2 vols, 8vo, 10s 6d bound
78 Linden on the Waters of Landrindod, 5s bound
79 Liger on the Gout, 5s bound
80 London Practice of Phyfic, 5s bound
81 Lobb's Practice of Phyfic, 2 vols, 10s
82 Lommius on Fevers, 5s bound
83 Linde on the Scurvy, 6s bound
84 Laws relating to Phyficians, Surgeons, aud Apothecaries, digefted by T. Cunningham, 2s 6d
85 Monro's (Alex. jun.) Obfervations on the Structure and Functions of the Nervous Syftem, with 55 Copper-Plate Tables, folio, 2l 12s 6d in boards
86 ——— Phyfiology of Fifhes, 2l 2s
87 Monro (D) on the Difeafes of Military Hofpitals, 2 vols, 10s bound
88 ——— on Mineral Waters, 2 vols, 12s
89 ——— Prælectiones Medicæ, 2s 6d fewed
90 Monro's (Alex. fen.) comparative Anatomy, 2s 6d, fewed
91 ——— Anatomy, with the above, 5s bound
92 ——— Works, 4to 1l 4s boards
93 Motherby's New Medical Dictionary, or General Repofitory of Phyfic; containing an Explanation of the Terms, and an Account of the Difcoveries and Improvements in Anatomy, Phyfiology, Phyfic, Surgery, Materia Medica, Pharmacy, &c. with copper plates, folio, *2d edit. enlarged*, 2l 2s bound
94 Morgagni on the Seat and Caufes of Difeafes, tranflated by Alexander 3 vols, 4to 1l 11s 6d
95 Macbride's Practice of Phyfic, 4to, 1l 1s in boards
96 ——— Effays, 5s bound
97 Millar's Obfervations on the prevailing Difeafes in Great Britain, 4to, 12s boards
98 ——— Obfervations on the Management of the prevailing Difeafes in Great Britain, particularly in the Army and Navy, 4to 16s boards
99 ——— on Antimony, 8vo 2s
100 ——— Difcourfe on the Duty of a Phyfician, 4to, 1s 6d
101 ——— Obfervations on the Practice of the Weftminfter Difpenfary, 4to 5s fewed
102 Mead Monita & Precepta Medica, by Wintringham, 2 vols 10s bound
103 Medical Obfervations and Inquiries, 6 vols, 2l 2s bound
104 Medical Communications by the Society for promoting Medical Knowledge, 6s
105 Mills on the Management and Difeafes of Cattle, 7s bound
106 Medical Regifter, or General Lift of Practitioners in Phyfic and Surgery, both in Town and Country, and feveral Parts abroad, 4s 6d boards, 1783
107 Newman's Chemiftry, by Lewis 2 vols, 8vo, 12s, bound
108 Nicholfon's Introduction to Natural Philofophy, with Plates, 2 vols, 8vo, 12s boards
109 Obfervations on the Character and Conduct of a Phyfician, in twenty Letters, 8vo, 2s 6d
110 Pearfon's Principles of Surgery, for Students, Part I, 8vo, 5s boards
111 Prieftley's Hiftory of Electricity, 4to, 5th edit. with a Continuation now in the Prefs

112 Prieſtley's Introduction to the Study of Electricity, 2s 6d
113 —— Experiments and Obſervations on different Kinds of Air, 6 vols, 8vo, 1l 16s. in boards
114 Pott's Works, complete, 3 vols, 8vo, 1l 4s bound

The following Pieces of Mr. Pott's may be had ſeparate.

115 I. ——— the Fiſtula in Ano, 2s 6d ſewed
116 II. ——— the Fiſtula Lachrymalis of the Eye 1s 6d
117 III. Chirurgical Obſervations relative to the Cataract, the Polypus of the Noſe, the Cancer of the Scrotum, the different Kinds of Ruptures, and the Mortification of the Toes and Feet, 3s ſewed
118 V. Remarks on Amputation—On the Palſy of the lower Limbs, 1s 6d
119 Percival's Eſſays, Medical and Experimental, 3 vols, 8vo, a new edition is now (January 1788) preparing for the Preſs in 2 vols, 8vo, with conſiderable improvements
120 Percival's Obſervations and Experiments on the Poiſon of Lead, 2s
121 Park's Account of a new Method of treating Diſeaſes of the Joints of the Knee and Elbow, in a Letter to Mr. Pott, 1s 6d
122 Pharmacopœia Londinenſis, 4to, new edit. 1788, 9s in boards
123 Pringle on the Diſeaſes of the Army
124 Pearſon's Experiments on Buxton Waters, 2 vols, 10s bound
125 Quincy's Diſpenſatory, 7s bound
126 Rigby's Obſervations on Uterine Hæmorrhages, with Caſes, 3d edit. enlarged 3s
127 ——— on the Red Peruvian Bark, 2s
128 ——— on the Theory of the Production of Animal Heat, and on its Application in the Treatment of cutaneous Eruptions, and other Diſeaſes of the Surface, 4s
129 ——— Chemical Experiments on Sugar, 2s
130 Ruſſel's Oeconomy of Nature, 8vo, 5s
131 Smith's Formulæ, Medicamentorum, or Compendium of the Practice of Phyſic, 5s in boards
132 Sims's Obſervations on Fevers, and other Epidemic Diſorders, 2d edit. 5s bound
133 —— Diſcourſe on the beſt Method of proſecuting Medical Enquiries, read before the Medical Society, 2d edit. 2s
134 Smellie's Midwifery, 3 vols, 8vo, with plates, 1l 1s bound
135 Simmons's Elements of Anatomy, and the Animal Oeconomy, 2d edit. 6s bound
136 ——— Anatomy of the Human Body, vol 1, 6s boards
137 ——— on the Gonorrhœa, 2d edit. 1s 6d
138 ——— Account of the Tænia, 2d edit. 2s 6d
139 Saunders on the ſuperior Efficacy of the Red Peruvian Bark in the Cure of Agues and other Fevers, 4th edit. 3s
140 ——— Tranſlation of Plenck on Mercury, 2s
141 Saviard's Surgery, 5s
142 Syſtem of Anatomy, 3 vols, 18s in boards
143 Smellie's Theſaurus Medicus, 4 vols, 1l 5s boards
144 Stark's Works, conſiſting of Clynical and Anatomical Obſervations, and Experiments on Diet, with Plates, 4to, 10s 6d in boards
145 Scheele's Experiments on Air and Fire, with an Introduction by Bergman: tranſlated by Dr. Forſter, with Notes by Mr. Kirwan, and a Letter from Dr. Prieſtley, 3s 6d ſewed
146 Swediaur on obſtinate Venereal Complaints, 4s
147 Theobald's Diſpenſatory, compiled for the Uſe of the Army, 3s bound
148 Watkinſon's Examination of a Charge brought againſt Inoculation by De Haen, Raſt, Dimſdale, and others, 1s 6d
149 Willan's Obſervations on the Sulphur Water at Croft, near Darlington, 1s 6d
150 Warner's Caſes in Surgery, 4th Edition, with additional Caſes, 6s in boards
151 Webſter's Medicinæ Praxeos Syſtema, 3 vols, 13s in boards
152 White on the Diſeaſes of Lying-in Women, 6s in boards
153 Wilſon's Medical Reſearches, on the Blood, and on Hyſterics, 4s ſewed
154 Zimmerman on Experience in Phyſic, 2 vols, 8vo, 12s bound

www.ingramcontent.com/pod-product-compliance
Lightning Source LLC
Chambersburg PA
CBHW021657210326
41599CB00013B/1445

9783337042554